2016 中国建筑院校境外交流优秀作业集

全国高等学校建筑学专业教育评估委员会
中国建筑学会建筑教育评估分会

中国建筑工业出版社

中国建筑学会建筑教育评估分会

中国建筑学会建筑教育评估分会在住房和城乡建设部人事司的直接领导和推动下，经中国科学技术协会、民政部批准，于2012年10月在北京成立。

中国建筑学会建筑教育评估分会的主要任务是与全国高等学校建筑学专业教育评估委员会、全国高等学校建筑学学科专业教学指导委员会协调工作，共同搭建一个活动平台，开展国际及港澳台地区学术交流与合作，参与建筑学专业教育国际互认《堪培拉建筑教育协议》有关活动，维护我国公民参加国外建筑师注册时享有与本国建筑专业教育背景同等地位的权利；跟踪建筑专业教育和学科的国际、国内发展趋势，动态分析建筑专业教育现状，努力促进建筑专业教育学术繁荣和技术进步，为中国建筑院校学术研讨和交流提供服务。

建筑教育评估分会成立以来，召开了五次理事会，健全了组织机构和工作制度，承担了堪培拉建筑教育协议中国秘书处工作，印制了《堪培拉建筑学教育协议》文件汇编，成功举办了四次中国建筑院校学生境外交流作业展并于2013、2014、2015年分别出版《中国建筑院校学生境外交流作业集》，积极组织策划了中英建筑院校学生工作坊、海峡两岸建筑院校学术交流和中韩建筑院校学生交流工作坊等教学交流活动。

建筑教育评估分会秘书处设在中国建筑学会国际部，不定期出刊中国建筑学会建筑教育评估分会简讯。

附：中国建筑学会建筑教育评估分会第一届理事会名单

理 事 长：朱文一（清华大学建筑学院）
副理事长：王建国（东南大学建筑学院）、仲德崑（深圳大学建筑与城规学院）
秘 书 长：张百平（中国建筑学会）
副秘书长：王柏峰（住房和城乡建设部）、王晓京（中国建筑学会）
常务理事：（按姓氏笔画排序）
丁沃沃（南京大学建筑与城市规划学院）、王建国（东南大学建筑学院）、孔宇航（天津大学建筑学院）、卢峰（重庆大学建筑城规学院）、朱文一（清华大学建筑学院）、仲德崑（深圳大学建筑与城市规划学院）、刘克成（西安建筑科技大学建筑学院）、刘临安（北京建筑大学建筑与城市规划学院）、孙一民（华南理工大学建筑学院）、孙澄（哈尔滨工业大学建筑学院）、李早（合肥工业大学建筑与艺术学院）、李保峰（华中科技大学建筑与城市规划学院）、吴长福（同济大学建筑与城市规划学院）、吴越（浙江大学建筑工程学院建筑系）、沈中伟（西南交通大学建筑与设计学院）、张伶伶（沈阳建筑大学建筑学院）、范悦（大连理工大学建筑与艺术学院）、单军（清华大学建筑学院）、赵继龙（山东建筑大学建筑城规学院）、韩冬青（东南大学建筑学院）、魏春雨（湖南大学建筑学院）

理　　事：（按姓氏笔画排序）
丁沃沃（南京大学建筑与城市规划学院）、于文波（浙江工业大学建筑工程学院建筑系）、马明（内蒙古科技大学建筑与土木工程学院）、王晓（武汉理工大学土木工程与建筑学院）、王薇（河南工业大学土木建筑学院建筑系）、王万江（新疆大学建筑工程学院）、王建国（东南大学建筑学院）、王绍森（厦门大学建筑与土木工程学院）、孔宇航（天津大学建筑学院）、石磊（中南大学建筑与艺术学院）、卢峰（重庆大学建筑城规学院）、吕品晶（中央美术学院建筑学院）、朱文一（清华大学建筑学院）、朱雪梅（广东工业大学建筑与城市规划学院）、仲德崑（深圳大学建筑与城规学院）、刘仁义（安徽建筑大学建筑与规划学院）、刘克成（西安建筑科技大学建筑学院）、刘临安（北京建筑大学建筑与城市规划学院）、关瑞明（福州大学建筑学院）、孙一民（华南理工大学建筑学院）、孙良（中国矿业大学建筑与设计学院）、孙澄（哈尔滨工业大学建筑学院）、严龙华（福建工程学院建筑与城乡规划学院）、李之吉（吉林建筑大学建筑与规划学院）、李早（合肥工业大学建筑与艺术学院）、李保峰（华中科技大学建筑与城市规划学院）、杨卫丽（西北工业大学力学与土木建筑学院建筑系）、吴长福（同济大学建筑与城市规划学院）、吴越（浙江大学建筑工程学院建筑学系）、沈中伟（西南交通大学建筑与设计学院）、张伶伶（沈阳建筑大学建筑与规划学院）、张建涛（郑州大学建筑学院）、张健（上海交通大学船舶海洋与建筑工程学院建筑学系、上海交通大学建筑设计及景观环境研究所）、陈洋（西安交通大学人居环境与建筑工程学院建筑学系）、陈晓卫（河北工程大学建筑与艺术学院）、武联（长安大学建筑学院）、范悦（大连理工大学建筑与艺术学院）、林耕（天津城建大学建筑学院）、周波（四川大学建筑与环境学院）、单军（清华大学建筑学院）、孟聪龄（太原理工大学建筑与土木工程学院）、赵继龙（山东建筑大学建筑城规学院）、郝赤彪（青岛理工大学建筑学院）、胡振宇（南京工业大学建筑学院）、饶小军（深圳大学建筑与城规学院）、费迎庆（华侨大学建筑学院）、姚赯（南昌大学建筑工程学院建筑系）、贾东（北方工业大学建筑与艺术学院）、贾晓浒（内蒙古工业大学建筑学院）、夏健（苏州科技大学建筑与城市规划学院、苏州国家历史文化名城保护研究院）、夏海山（北京交通大学建筑与艺术学院）、龚兆先（广州大学建筑与城市规划学院）、隋杰礼（烟台大学建筑学院）、韩冬青（东南大学建筑学院）、程世丹（武汉大学城市设计学院）、舒平（河北工业大学建筑与艺术设计学院）、翟辉（昆明理工大学建筑与城市规划学院）、戴俭（北京工业大学建筑与城市规划学院）、魏春雨（湖南大学建筑学院）

秘 书 处：陈玲（中国建筑学会）、周政旭（清华大学建筑学院）、赵建彤（清华大学建筑学院）、商谦（清华大学建筑学院）

变革中的重新思考

《2016年中国建筑院校境外交流学生优秀作业集》序

2016年3月19日至20日，中国建筑学会建筑教育评估分会2016年年会暨第一届五次理事会在哈尔滨工业大学建筑学院召开，本次会议由中国建筑学会建筑教育评估分会主办，哈尔滨工业大学建筑学院承办。住房和城乡建设部人事司副巡视员赵琦、中国建筑学会理事长修龙、常务副秘书长张百平及56个理事单位的理事及代表等约170余人出席了会议，共同就"专业教育评估与设计课程教学"这一主题展开了深入的研讨。在此同时，"2016年中国建筑院校境外交流优秀作业展"及优秀作业的评选也如期展开。

本次学生作业展览及评选针对2014—2015学年度（2014年9月至2015年8月）本科生或研究生在与境外建筑院校交流中所完成的（联合）设计作业。截至2016年3月10日，共收到来自国内41所建筑院校的251份作业。参与院校数量与作品数量较上一届有了较大幅度的提高。经过评选前的技术审查，2份作业因不符合规范被取消参评资格，最终参评作业249份。

评选工作在哈尔滨工业大学建筑学院体育馆举行，分会理事长朱文一教授主持了评选。优秀作业评审委员会由中国建筑学会副理事长周畅，建筑教育评估分会理事长朱文一，副理事长王建国、仲德崑，常务理事孔宇航、刘临安、孙澄、孙一民、李保峰、赵继龙、张伶伶、丁沃沃，哈尔滨工业大学建筑学院、哈尔滨工业大学建筑设计研究院院长梅洪元，哈尔滨方舟工程设计咨询有限公司董事长刘远孝，中国建筑东北设计研究院有限公司总建筑师赵成中，台北中华全球建筑学人交流协会理事长、淡江大学建筑系教授陆金雄，西安建筑科技大学建筑学院副院长李昊，同济大学建筑与城市规划学院副院长黄一如，重庆大学建筑城规学院副院长卢峰，浙江大学建筑工程学院建筑系主任吴越和大连理工大学建筑与艺术学院院长范悦共21位成员组成。评审工作经过四轮的讨论和投票，第一轮采用淘汰表决，第二、三、四轮采用晋级表决，最终选出94份"2016年中国建筑院校境外交流学生优秀作业"，获奖学生及指导教师由中国建筑学会建筑教育评估分会分别颁发"2016年中国建筑院校学生境外交流优秀作业证书"及"指导教师证书"。作品集结出版，是为了让国内更多的建筑院校和读者了解、学习和参考。

随着世界各国之间相互依存、相互影响不断增强，全球地缘政治的变化更加迅速。我们所面临的社会问题和挑战是全球性的，然而解决方案却因国家和地区的不同而具有各自特点。中国建筑院校与境外建筑院校、设计机构的交流恰恰为我们提供了观察、探讨这一问题的机会。纵观本次送展参评的作品，突出呈现出对以下三个问题的关注：

社会变革中的城市姿态

现代化和民众对工业化福利社会的诉求对建筑师所追求的公平提出了全新的判断标准。在当今世界，城市中财富与贫困并存；一些国家面临人口严重老龄化问题，另一些国家则需要控制高出生率。我国更是同时面临着人口规模大和老龄化的双重挑战。本次多份参评作业关注弱势或特殊群体，为老龄化以及大众生活而设计，如广州大学—香港中文大学—英国格拉斯哥艺术学院的联合工作坊"抑郁症社区康复中心"、"城中村中的新市场"和"自闭症社区康复中心"，湖南大学—斯洛文尼亚卢布尔雅那大学合作的"光炫之城——长沙历史街区中的社区图书馆"，南京工业大学—美国堪萨斯大学的联合毕业设计"年龄混合型的老年居住、健康、医疗、康复社区规划与建筑设计"，哈尔滨工业大学—西班牙BAUM建筑设计事务所的联合设计"泡泡城市"、"中国城市建筑工地预制工人住宿"和"老年人日托中心设计"，山东建筑大学—新西兰国立理工学院的合作设计"织补·起航——移民社区微型功能集合体"，福建工程学院—台湾科技大学的联合毕业设计"明眸心塾——基于新理念下的视障教育学校设计"等作品。这些作品探索了面对城市中人的精神健康下降、空巢老人增加、外来务工人员生活质量低下、视障孩子接受教育困难等社会问题，以及在解决这些问题过程中建筑师的角色定位和建筑应对之道。通过这些作品，可以看出我国建筑院校对学生进行知识和技能培养的同时，还注重社会责任和人文关怀的培养，引导学生在设计中对社会问题进行深刻的思考和探索。

城市更新中的既有和传统

高速行进中的城市化进程，古建筑被推倒、历史街区被破坏、废弃厂区成为城市危险地带、乡村失去了特色，这对建筑学人提出了严峻挑战。本次参评作业中有相当一部分作业针对地区性城市发展的特殊问题和矛盾，探讨了建筑解决之道，如哈尔滨工业大学—台湾文化大学的开放设计"猴硐综合体—基于产业活化的选煤厂改造"，西安建筑科技大学—意大利米兰理工大学的联合工作营"唐轴线小雁塔地段城市设计"，湖南大学—捷克技术大学的合作设计"长沙滨江天伦造纸厂更新改造联合设计"，东南大学—维也纳工业大学合作的"南京大行宫碑亭巷旧居住区更新改造设计"，天津大学—意大利罗马大学合作的"陶片山山墙区域城市设计及旧建筑加改建设计"，重庆大学—日本早稻田大学—香港大学的合作设计"山地田园综合体——松阳塘后村村落更新设计"、"织补——基于历史文脉重塑的松阳老街更新设计"，太原理工大学—荷兰Karel事务所的联合设计"越陌度阡"，华侨大学—澳门土地工务运输局—澳门文化局的合作项目"澳门十月初五街改造与更新设计"，北京建筑大学—德国柏林工业大学的联合设计"保福寺地区城市更新项目"，华中科技大学—德国慕尼黑大学的联合设计"花楼街改造——谦卑的抵抗"，大连理工大学—德国达姆施塔特大学的合作设计"Pyeonghwa市场改造"等作品。这些作品不仅探索了历史街区保护、城市棕地改造等物质层面的问题，更探究了传统生活方式留存、城市记忆延续等精神层面的问题，展现出建筑学人面对城市既有建筑和传统文化所作出的思考和应对。

概念革新下的城市设计

城市的概念正在发生变化：城市与乡村的景象开始产生重叠，建设区域逐渐与景观和自然融合。如何重新定义城市的面貌？城市的未来会是怎样的图景？本次参评作业中有关城市设计的题目占据了相当大的比例，如清华大学—美国耶鲁大学的联合设计"天津滨海新区新河船厂城市设计"，哈尔滨工业大学—英国谢菲尔德大学的联合设计"复合进化论——自混沌至有机的艺术社区的自发成长"，东南大学—日本京都大学—日本三菱地所的合作设计"南京地铁马群站城市设计"，山东建筑大学—澳大利亚昆士兰理工大学的联合设计"适合步行的城市——布里斯班Westend城市设计"，浙江大学—西澳大利亚大学的专题化设计"万花城——人与生态的城市设计"，华南理工大学—意大利都灵理工大学的联合工作坊"T.I.T工业创意园城市设计策略探究"，中国矿业大学—新西兰奥克兰大学的联合设计"都市森林——徐州韩山东路地块城市设计"，清华大学—新加坡国立大学的联合工作坊"优联都市，活力港湾——新加坡裕廊工业区总体城市设计"等作品。这些作品均是对城市过去的总结与反思、对现在问题的关注，并探索提出面向未来的解决方案。

我国建筑院校设计教学的境外合作交流发展到今天，已经发展成为常态的、多形式的、跨专业的、研究型的特色活动，不仅是教学上的拓展，还发展成为学生出国深造、教师合作研究的重要平台。在境外教学或设计机构的合作交流中，师生们在时代和社会的变革背景下重新探讨城市的定义、建筑的姿态、人与城市的关系，重新思考建筑学的范畴和定位、建筑学人的责任和角色，所有这些对于培养具有社会责任感、善于思考、勇于突破的建筑人才具有着重要的意义。

哈尔滨工业大学建筑学院　梅洪元
哈尔滨工业大学建筑学院　孙　澄
2016年8月

目　录

变革中的重新思考

1. 猴硐综合体——基于产业活化的选煤厂改造
   Rui San New Established SOHO ································· 10
2. 学术景观——剑桥水边仓库图书馆改造
   Academic landscape: Regenerate a Riverside Warehouse as an Emblem of Active Learning ································· 12
3. 天津滨海新区新河船厂城市设计
   Xinhe Shipyard Renovation, Tianjin, China ································· 14
4. 泡泡城市
   Bubble City ································· 16
5. 之间——济南商埠区中山公园东片区综合体设计
   IN-Between: Urban Hybird Building Design ································· 18
6. 21世纪的图书馆——台中创意设计中心
   Taizhong Creative & Design Center ································· 20
7. 中渭桥遗址博物馆设计
   Museum Design of Wei Bridge remains of Qin and Han Dynasty ································· 22
8. 抑郁症社区康复中心
   Rehabilitation Neighbourhood ································· 24
9. 泉景——泉水博物馆设计
   Springscape—Spring Theme Museum Design ································· 26
10. 双面的波哥大
    Bogota in Duality ································· 28
11. 灯塔
    Leucht Turm ································· 30
12. 目的地的兴盛——云南抚仙湖规划建筑设计
    Yunnan Fuxian Lake Resort Design ································· 32
13. 回溯唐的未来——唐轴线小雁塔地段城市设计
    Back to the Future of Tang Dynasty ································· 34
14. 织补 · 起航——移民社区微型功能集合体
    Darning Set Sail ································· 36
15. 长沙滨江天伦造纸厂更新改造联合设计
    Regional Architecture—Industrial Reconstraction and Renewal Design of Changsha Papermaking Factory ································· 38
16. 南京大行宫碑亭巷旧居住区更新改造设计
    The Renovation of Nanjing Beiting Lane Residential District ································· 40
17. 在历史中徘徊
    Wandering through the History ································· 42

18. 山—海—人的相遇——温哥华卑诗省沿海高速公路木休闲站设计
    Where Rivers, Mountains and People Meet: Wood Pavillion in Highway Stop, Vancouver, BC 44
19. 台北青年旅馆设计
    The Green Track 46
20. 悬浮森林——滨江新城厂房改扩建之图文媒体中心
    Floating Blocks-Library and Information Centor of the community 48
21. 光炫之城——长沙历史街区中的社区图书馆
    City of Blinding Lights: Community Library in Historical District of Changsha 50
22. 山地田园综合体——松阳塘后村村落更新设计
    Mix of Mountain and Country Side: Return to a Pastoral Songyang 52
23. 复合进化论——自混沌至有机的艺术社区的自发成长
    The Theory of Hybrid Evolution: The Artists Community's Sponteneous Growth from Chaos to Invisible Order 54
24. 我的微纽约——位于中央车站的垂直市场大楼
    My Tiny New York: A Vertical Market Tower at Grand Central Terminal 56
25. 城市嘉年华
    Urban Carnival 58
26. 都市驿站——垂直建筑系馆设计
    City Courier station-Vertical Campus Design 60
27. 垂直价值激发器
    Rediscover the Value High up in New York 62
28. "织补"——基于历史文脉重塑的松阳老街更新设计
    Weave Repair update 64
29. 旧城改造
    fidenzal 66
30. 序列·传承——唐轴线小雁塔地段城市设计
    Sequence-Inheritied 68
31. 一方天地——西安渭桥遗址区博物馆设计
    One Square World: Xi'an Wei Bridge Heritage Site Museum 70
32. 解锁——过渡性住宅
    Transition-Intermediary Home 72
33. 年龄混合型的老年居住、健康、医疗、康复社区规划与建筑设计
    Inter-generational Community: Serior living, Health and Wellness, Rehab, Clinic 74
34. 城脉
    Pulse of Old Town 76
35. 城市建筑工地预制工人住宿
    City Green "Wall" 78
36. 共生——怡园历史街区更新城市及建筑设计
    Symbiosis: Urban Design and Architecture Design of Joyous Garden Historical District 80

37. 扩散
    Diffusion ·· 82
38. 应变，随时而变
    Change over Time ·· 84
39. 南京地铁马群站城市设计
    Nanjing Subway Station Urban Design ··· 86
40. 传承·规矩——唐轴线小雁塔地段城市设计
    Inheritage and Rules-City Design of the Xiaoyan Pagoda Area in the Axis of
    Tang Dynasty ·· 88
41. 行走在丝路之上
    Walking on the Silk Road ·· 90
42. 步行城市——布里斯班韦斯滕德城市设计
    Walkable City ·· 92
43. 隙光——意大利普拉托考古遗址博物馆设计
    Gap and Light: Prato Museum Design ··· 94
44. 越陌度阡
    Walking through the Streets ··· 96
45. 互动——剑桥大学学院制下公共空间的分析与设计
    Interaction: Analysis of Common Spaces in the Colleges in the University of
    Cambridge ·· 98
46. 万花城——人与生态的城市设计
    Kaleido City ···100
47. 种植·培养·建造
    Plant·Cultivate·Generate ···102
48. 地球村
    The Global Village ···104
49. 保福寺地区城市更新项目（一）
    Urban Renewal Project of Baofusi District(1) ······································106
50. 保福寺地区城市更新项目（二）
    Urban Renawal Project of Baofusi District(2) ······································108
51. 保福寺地区城市更新项目（三）
    Urban Renewal Project of Baofusi District (3) ·····································110
52. 博登湖畔市政厅设计
    Rathaus am Bodensee ··112
53. 青年旅馆设计
    Design of Youth Hostel ··114
54. 和平市场改造计划
    Pyeonghwa Market Regeneration ···116
55. 社区建筑——东南大学书院设计
    Community Building ···118
56. 住区设计
    Adaptive Residence ··120

57. 波士顿新区音乐广场设计
    Boston Music Playground 122

58. 年龄混合型的老年居住、健康、医疗、康复社区规划与建筑设计
    Inter-generational Community: Senior living, Health and Wellness,
    Rehab, Clinic 124

59. 转变·演绎——唐轴线朱雀门顺城巷地段城市设计
    City Wall Lane Area Transformation and Connection Design 126

60. 行走—延续——西安渭桥遗址区博物馆设计
    Continuation by Walk 128

61. 啤酒厂改造
    Reconstruction of Brewery 130

62. T.I.T. 工业创意园城市设计策略探究
    T.I.T. Urban Design Strategies Reserch 132

63. 活力台北——青年文创 SOHO 集合住宅设计
    Co-housing 134

64. 点·聚生活
    Attitude in Our Culture Area 136

65. 社区活动中心
    Rehabilitation Neighbourhood: for the Community Center 138

66. 自闭症社区康复中心
    Rohab-center for Autism 140

67. 城中村中的新市场
    A Route to Harmony through Market 142

68. 伊特鲁里亚的复兴——普拉托 Gonfient 地区遗址公园设计
    Reborn Etruscan 144

69. 缓缓归——都市集合住宅设计
    Walking Home Slowly: Design of Urban Collective Housing 146

70. 唤醒历史——古罗马西南角城墙区更新
    Wandering through the History 148

71. 垂直终点站
    Vertical Terminal 150

72. 明眸心塾——基于新理念下的视障教育学校设计
    Healing Sight Healing Soul: Based on New Concept of Special School Design ... 152

73. 冰城印象
    Ice City 154

74. 都市森林——徐州韩山东路地块城市设计
    Urban Forest 156

75. 农村社区中心
    Rural Service Center 158

76. 城市之下，山丘之上——居住+观光混合居住区设计
    Under City, Above Mount: Living+Visiting Residential Building 160

77. 优联都市，活力港湾——新加坡裕廊工业区总体城市设计
    Connected City & Vibrant Harbour: The Rejuvenation Design of Jurong Industrial District, Singapore 162
78. W.O.W. 大厅扩建
    An Expansion to W.O.W. Hall 164
79. 谦卑的抵抗——花楼街改造
    Humble Resistence—Hualou Street Restoration 166
80. 湖底巷——武汉中心老城区复兴
    The Sinking Alley 168
81. 天津城建大学体育馆设计
    The Stadium Design of Tianjin Chengjian University 170
82. 孵化新网络
    SOHO 2.0 172
83. 小建筑，微社区——浙江里庚村公共空间的激活与设计
    Micro-architecture 174
84. 西班牙传统手工艺博物馆设计
    Design of Spanish Craftwork Center 176
85. 美术馆
    Art Museum 178
86. 新集体
    The New Collective 180
87. 中国住宅设计文化的传承
    Wohnkulturen 182
88. 城市设计
    Urban Design 184
89. 旧桃·新符——住区复兴与住房保障
    Urban Rehabilitation and Housing Resocialization 186
90. 墨尔本联邦广场东侧地块城市设计（1）
    Developing Federation Square East, Melbourne（1） 188
91. 墨尔本联邦广场东侧地块城市设计（2）
    Developing Federation Square East, Melbourne（2） 190
92. 探寻花楼街——板车组
    Investigating Hualou Street 192
93. 自然环境——幼儿园设计
    Flowing in the Trees 194
94. 老年人日托中心设计
    Day Care Center Design 196

2016 年中国建筑院校境外交流优秀作业名单 198
中国建筑学会建筑教育评估分会第一届第五次理事会代表名单 205

后记

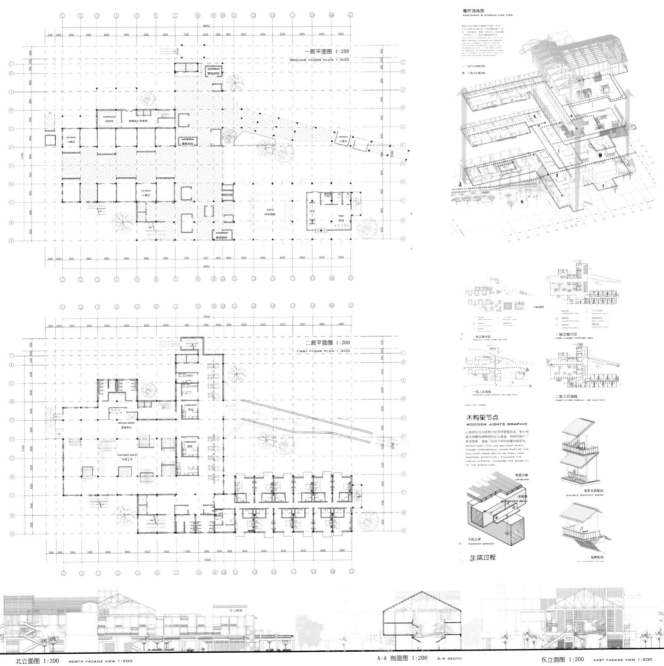

University: Harbin Institute of University
Designer: Zhang Xiangyu, Wei Na, Wang Yining
Tutor: Zhou Lijun, Yin Qing, Han Yanjun
Course: International Joint Design Studio
Finished Time: Apr., 2015
Exchange Institute: Chinese Culture University

First Prize

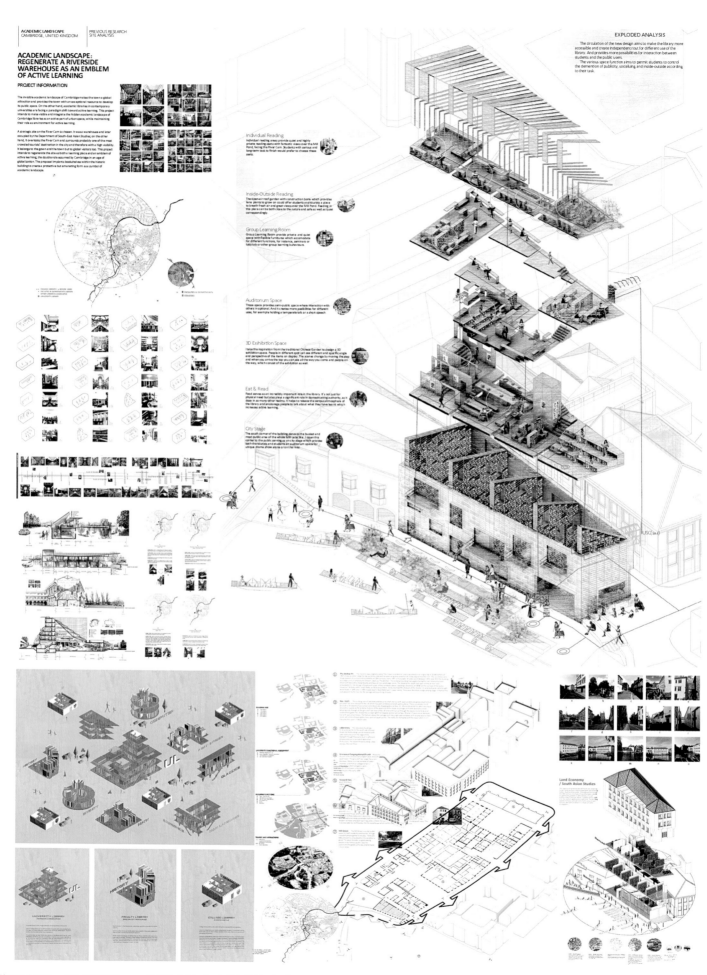

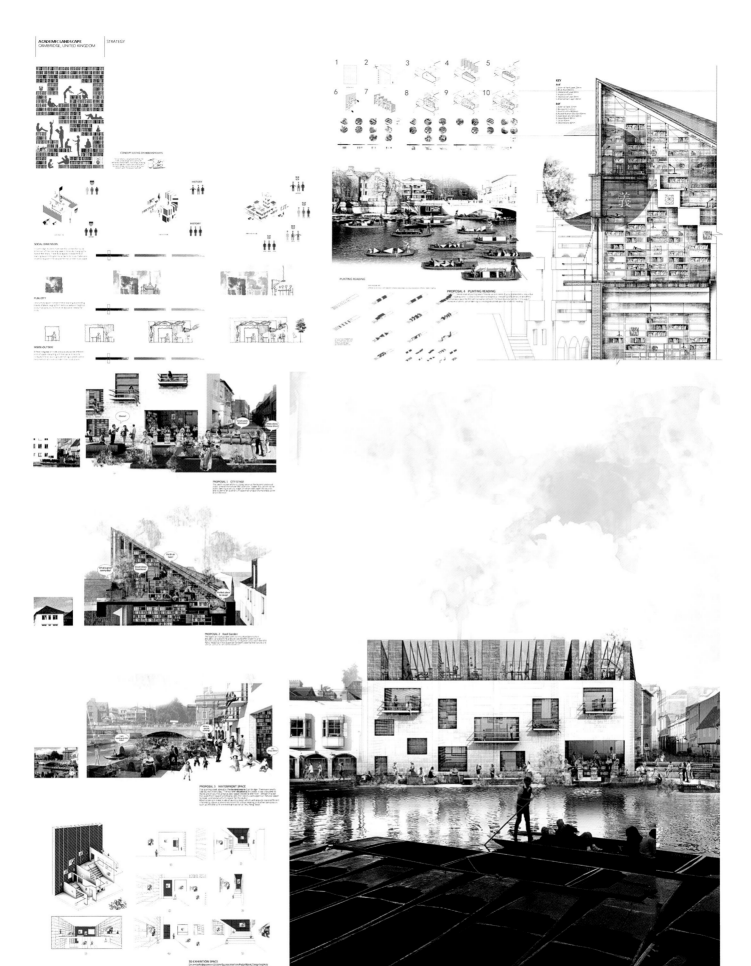

University: Nanjing University
Designer: Xi Hong
Tutor: Dou Pingping, Ingrid Schröder
Course: Nanjing-Cambridge University Joint Design Studio
Finished Time: Jul., 2015
Exchange Institute: School of Architecture, Cambridge University

First Prize

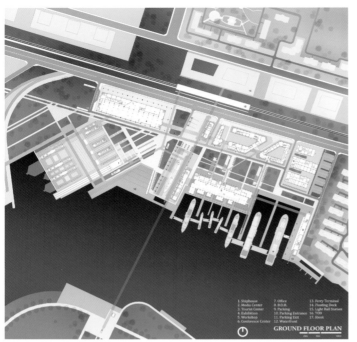

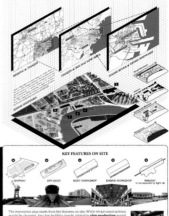

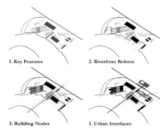
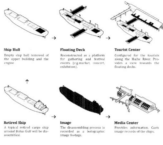

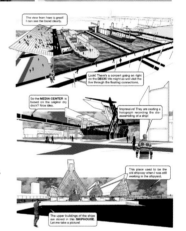
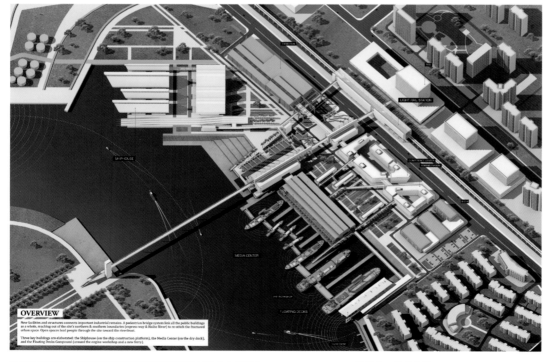
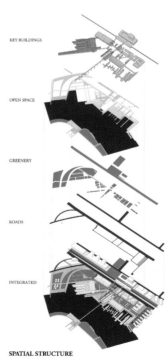

作品名称：天津滨海新区新河船厂城市设计
Xinhe Shipyard Renovation, Tianjin, China

二等奖

院校名：清华大学
设计人：卓信成，汪明全，蔡泽宇
指导教师：朱文一，刘健
课程名：2014—2015年度秋季学期研究生建筑设计studio
作业完成日期：2014年12月
对外交流对象：美国耶鲁大学建筑学院

SHIPHOUSE UNIT CROSS SECTION

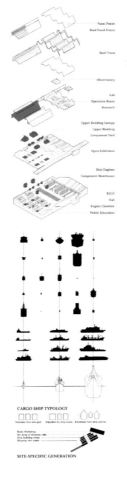

CARGO SHIP TYPOLOGY

SITE-SPECIFIC GENERATION

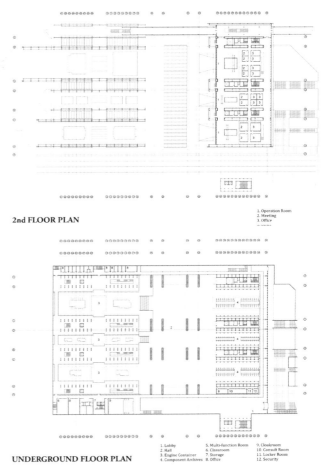

2nd FLOOR PLAN

1. Operation Room
2. Meeting
3. Office

UNDERGROUND FLOOR PLAN

1. Lobby
2. Hall
3. Engine Container
4. Component Archives
5. Multi-Function Room
6. Classroom
7. Storage
8. Office
9. Cloakroom
10. Consult Room
11. Locker Room
12. Security

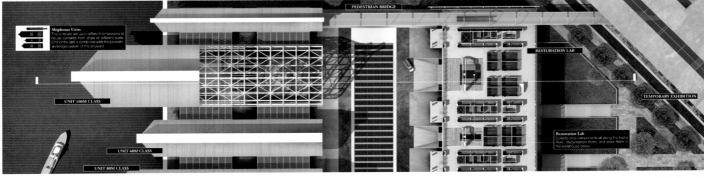

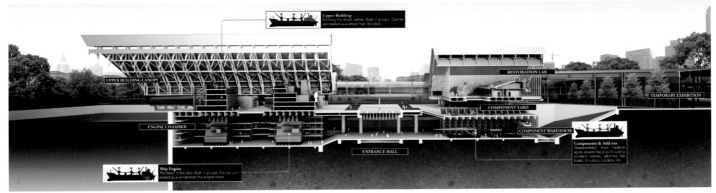

SHIPHOUSE UNIT PLAN & LONGITUDINAL SECTION

The canopy and the underground space act as containers of upper buildings and ship engines. According to the several classes of incoming ships, they are designed in different sizes and separated into units. The upper buildings and engines from retired ships, once decomposed, will be stored permanently.

The right half of the section presents the non-permanent part. The underground space as the component warehouse, stored decomposed fragments of retired ships. Upon it is the restoration lab. With a floor hatch, the lab allows its staff to have access to the newly imported ship components in the component yard in between. After study/restoration is completed, components are moved down to the warehouse.

Section Model Detail:

University: Tsing hua University
Designer: Cai Zeyu, Wang Minquan, Zhuo Xincheng
Tutor: Zhu Wenyi, Liu Jian
Course: Tsinghua-Yale Joint Studio
Finished Time: Dec., 2014
Exchange Institute: School of Architecture, Yale University

Second Prize

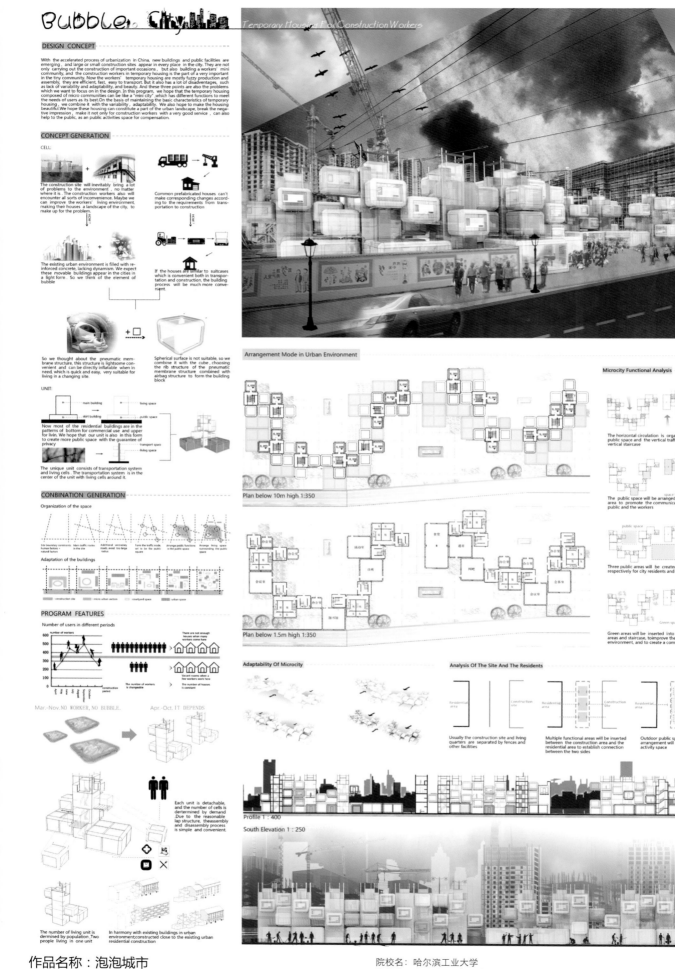

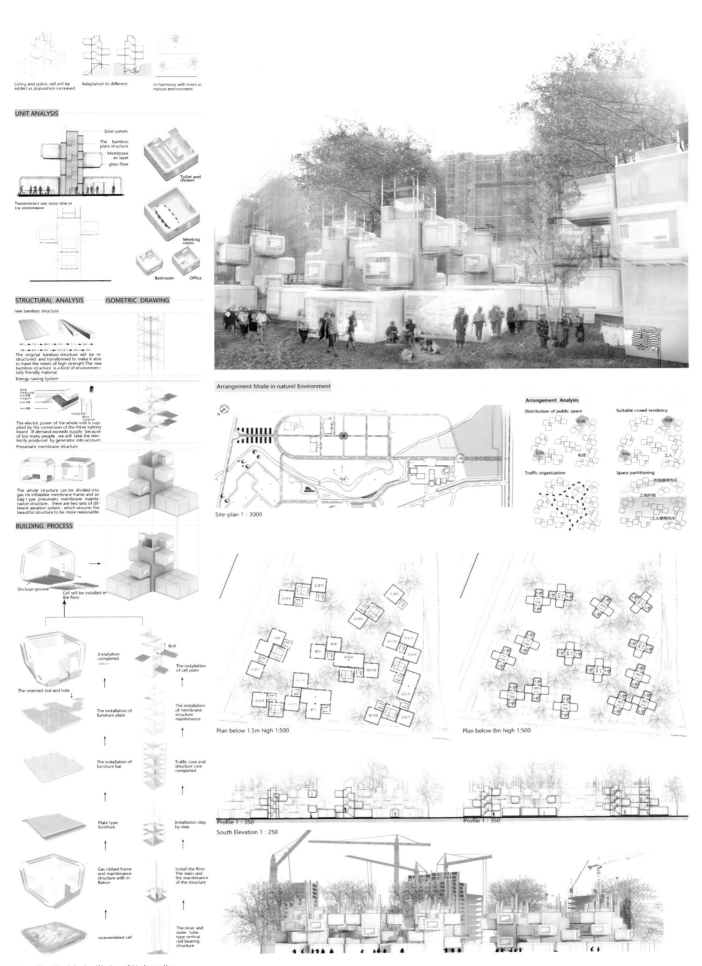

University: Harbin Institute of University
Designer: Bai Slyao, Wu Tong, Zhuang Mingchang, Kelly Charles
Tutor: Lian Fei, Tang Jiajun, Javier Caro Dominguez
Course: International Joint Design Studio
Finished Time: Jul., 2015
Exchange Institute: BAUM

Second Prize

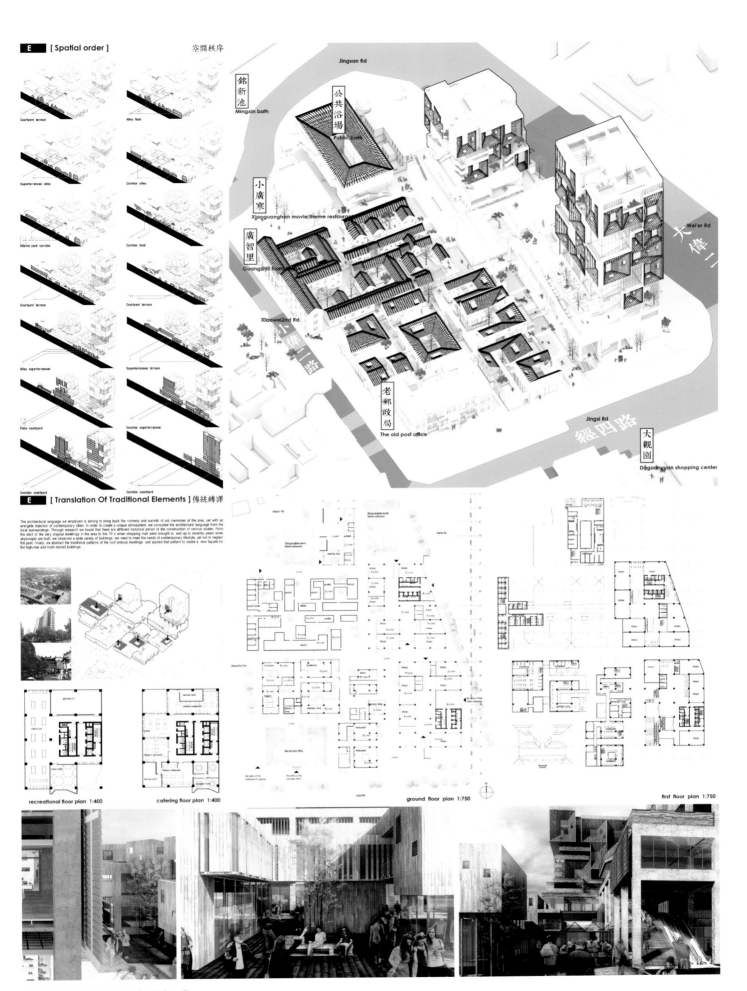

University: Shandong Jianzhu University
Designer: Li Peiru, Zhou Jifa, Wang Tianyuan
Tutor: Zhou Zhongkai, Liu Jianjun, Yvonne Wang, Paul Sanders
Course Name: Shandong Jianzhu University-Queensland University of Technology
Finished Time: Jun., 2015
Exchange Institute: Queensland University

Second Prize

21世纪的图书馆——台中创意设计中心
TAIZHONG CREATIVE & DESIGN CENTER

With the advent of the era of consultation, the library is modeling multi functional space for independent reading, multiple learning, integrated learning and lifelong learning. The role of library is starting from the passive stack role, gradually to become the Teaching Resource Center or learning resource center. It is no longer a place limited to find information while it has a certain innovation ability. All kinds of new media consultation, Science and technology have emerged in online databases, CD-ROM, electronic books, multimedia, hypermedia, Internet resources, which offers the library an important direction to Transform.

Today we are very easy to find information from the Internet, but not as predicted, the real figure of library is not only decline or disappear, but has more new vitality. As a cultural symbol, a symbol of civilization, library is irreplaceable. The library is accessible to more and more by being shared with the community, arranging for drama, music, movies and multimedia performances. Making full use of the public library is a significant contribution to the vitality of the city, also more likely for a library to be an important center of learning and social and meeting place.

Target:
1. definition of the library in the era of consulting.
2. logic and rationality of building system and organization
3. space content(usage, activity aspect) and space form interconnect
4. integration and corresponding with the context of the site
5. combination of architectural space as urban life context and urban public space
6. definition of the relationship between indoor and outdoor, the continuity of urban landscape and architectural space

21世纪的图书馆——台中创意设计中心 TAIZHONG CREATIVE & DESIGN CENTER 1

概念生成 / CONCEPT
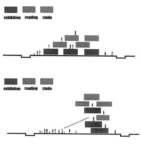
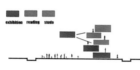

基地分析 / SITE ANALYSE

The site area of 3600 square meters, the building coverage ratio of 60%, the volume ratio of 220%. The site is for independent preset block, on the front of Sanmin street (25m), and Yizhong Street. The other side is Yucai Road (both 10M tunnel), which is located in core of the Yizhong Street business district. The other two sides of site stand NO.1 middle school of Taichung and Zhongyou department store, on which north is Yimin bussiness district.
There are school, department store and cram school in the area, hence it has long been teenagers (defined as 18 years of age ~25) gathering places. Thus in addition to special snacks, par restaurants, this area looks to sell fashion, bags, accessories, jewelry, stationery and other pop culture products.

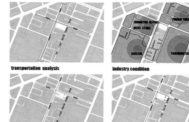
transportation analysis / industry condition
pedestrian flow analysis / path and brt station

This area is the high density region of crowds activities and traffic. The intention of site selection of TCDC to be here is to not just integrate with surrounding activities, also hope that in addition to business, can inject more activities in the city, stimulate and support, to increase the quality and ability of cultural and creative products.

residential space design / commercial space design

education space design / free space design

资—资源/资料/书籍/商品/Z信息
Info–material/resources–books/commodity/information
途—路径/道路/流线
Path–Path of site –the road / flown line
主体空间—资源中心/学校/图书馆/被管理空间
The main space—resource center / school / library /singlefunction space
剩余空间—无差别开放空间/无管理
Free space–indistinctive open space
中心—各种不同性质的场域交汇地
The center - intersection of fields with different property

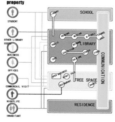

基地分析 / BASE ANALYSE

Shuttling in the site, differing the unified style of the contiguous storefront in the mainland, every building here is unique. The four-to-five-storey finales are with 4-5 meters face width, and the narrow lane is full with various boothes. The rough overline bridge and gallery bond visitors from around the world. Space that highly congested and placement of various mobile business form make the space rich and fulfilling.
At sunset, street function extending and to bear a variety of mobile vendors, tumult of the night market coming quietly. The not spacious land brings me may not shock, but is difficult to forget. Crowded but productive, romantic but practical, disorderly but organised. It is seemingly complete, but still lacks a place, whereas people could breathe freely in the crowded land!

功能分区 / FUNCTION DIVISION

public / transport
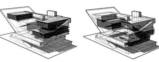
reading / exhibition

auxiliary / studio

配置图 / SITE PLAN

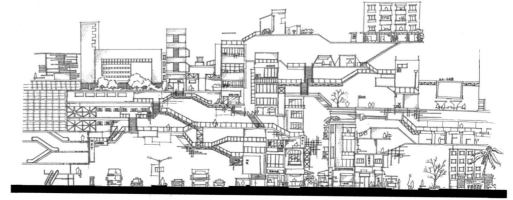

作品名称：21世纪的图书馆——台中创意设计中心
Taizhong Creative & Design Center

院校名：大连理工大学
设计人：赵玥
指导教师：罗曜辰，范悦，于辉
课程名称：建筑设计(VI)
作业完成日期：2015年06月
对外交流对象：德国达姆施塔特大学建筑系

二等奖

21世纪的图书馆——台中创意设计中心
TAIZHONG CREATIVE & DESIGN CENTER

随着咨询时代的来临，不再神圣高高在上，而透过网络成为流动的、开放的甚至睡手可得，图书馆作为一种特定的建筑类型有独特的运作模式，当人与书本与知识的关系产生变化，图书馆的空间也随之改变。随着咨询时代的来临，图书馆作为自主阅读学习、多元学习、统整学习和终身学习的多功能多样貌空间，图书馆的角色开始由消极被动的书库角色，渐转为教学资源中心或学习资源中心，各类新兴咨询媒体与科技纷纷出现，网上资料库、光碟资料库、电子图书、多媒体、超媒体、网际网路资源等，成为图书馆转型的重要方向。

今实体图书馆还未没落，反而具新的活力，图书馆作为文化符号、不可取代、图书馆的面向越来越广，可与社区共享，也可以安排戏剧、音乐、电影和多媒体表演活动，充分使用的公共图书馆对城市的活力有明显的贡献，更可能为重要的学习、社交中心场所。

目标：
1、咨询时代图书馆的再定义
2、建筑系统和组织的逻辑与合理性
3、空间内容（使用方式、活动样貌）与空间形式的相辅相成
4、与基地环境的整合及对应
5、建筑空间与都市的生活脉络和都市公共空间的相互结合
6、室内外关系的界定、都市景观和建筑空间的连续性

基地：
基地面积为3600平方米，建蔽率60%，容积率220%，基地预设为独立街廓，主要临三民路（25m）-中街，另一侧是青才北路（皆为10M道路），位于一中街商圈核心，两边各是中友百货和台中一中，住北是的居民商店，长期以来一直是青少年（聚集之场所，除了特色小吃、平价餐厅以外，此商圈以销售青少年流行文化的商品为；是活动密度高、人潮车潮热闹的区域。

结构系统分析
STRUCTURE ANAYLISE

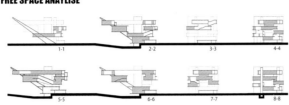

剩余空间分析
FREE SPACE ANAYLISE

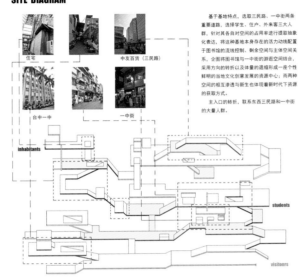

基地语汇
SITE DIAGRAM

基于基地特点，选取三民路、一中街两条重要道路，选择学生、住户、外客三大人群，针对其各自对空间的占用率进行提取抽象化表达，将这种基地本身有的活力动线配置于图书馆的流线控制。剩余空间与主体空间关系，企图将图书馆与一中街的游逛空间结合，采用方向的转折以及体量的退缩形成一座个性鲜明的当地文化蓬发展的资源中心。而两种空间的相互渗透与新生也体现着新时代下资源的获取方式。

主入口的转折 联系东西三民路和一中街的大量人群。

北立面图
NORTH ELEVATION

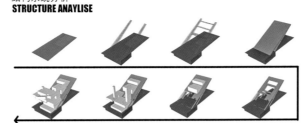

剖面图
SECTION

SECTION

SECTION B-B

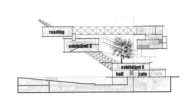

SECTION C-C

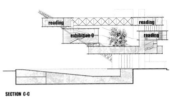

SECTION D-D

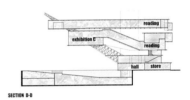

平面图
PLAN ANALYSIS

7F free space | auxiliary space | function space

6F free space | auxiliary space | function space

5F free space | auxiliary space | function space

4F free space | auxiliary space | function space

3F free space | auxiliary space | function space

2F free space | auxiliary space | function space

1F free space | auxiliary space | function space

21世纪的图书馆——台中创意设计中心　TAIZHONG CREATIVE & DESIGN CENTER 2

穿梭于这片基地之上，有别于大陆连片统一风格的临街店面，这里的每一栋建筑都各不相同，四五米的面宽穿穿升起一幢幢四五层高的小楼，窄窄的巷弄里也会搭满各种摊位，高高低低的天桥、廊道联系着闹来北住的游客，学生或是店主，在这里，空间的高度拥挤与各种移动式商业形式的置入，使空间丰富而充实，白日里，逛街客悠悠闲闲走过各具一格的小店，午休时成为一中学生的聚集地，日落时，承载着各式移动商贩，夜市的喧嚣悄然而至，在这里，城市剩余空间不断延伸……这片并不宽敞的土地，带给我的不是富裕，却是深刻的记忆，拥挤而丰富，无序却不混乱，这里看似完整，实则缺少一处空间，一片可以在这片拥挤暗闹，复杂多样的土地上"自由呼吸"的开敞空间。

方案伊始，考虑将基地肌理、街道立面和资讯时代图书馆功能的配置的新需求相结合，在两条看截然不同的线索下开头并进。寻找二者的契合点。在基地环境处理上，面对三民路一侧，延续主街上的商业界面，保证立面的整体性。面向一中街一侧配置入斜板，产生板上和板下两处室外开敞场地，为拥挤的基地和图书馆内部均提供了宽敞的"呼吸"空间。

斜板上承载起图书馆的主体功能，图书馆内空间配置延基地肌理，逶迤在一起。资讯时代，信息交换、资源共享，节奏快速的特点要求图书馆内空间进行变革，不同功能的空间联系却互不干扰。

同时，"斜板"强大的视觉冲击力和幕洞式的空间具有明显的人流集聚功能和信息吸纳力，至此，21世纪的台中创意中心概念完成。

University: Dalian University of Technology
Designer: ZhaoYue
Tutor: Luo Yaochen, Fan Yue, Yu Hui
Course Name : Architectural Design Studio(VI)
Finished Time: Jun., 2015
Exchange Institute: Technische Universität Darmstadt

Second Prize

中渭桥遗址博物馆设计
Museum design of Wei bridge remains of Qin and Han Dynasty

An Architectural Exploration of the Axis and the Process of the RETROSPECT

IDEA
After rigorous research of the Wei Bridge archeological site and its surroundings, we chose to highlight two main characteristics relevant to the history of both past and present. We want to give importance to the process of how the bridge was revealed. Since the bridge remains were discovered by local farmers, we kept the Thatch and wheat as a continued source in this area. The visitors will be able to become the "archeologists" themselves by experiencing the field when first entering the site, and later stumbling upon the museum.

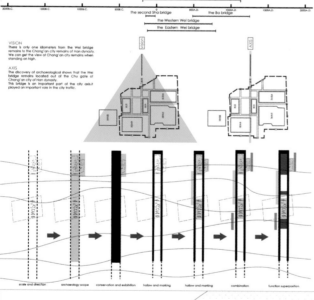

VISION
There is only one kilometers from the Wei bridge remains to the Chang'an city remains of Han dynasty. We can get the view of Chang'an city remains when standing on high.

AXIS
The discovery of archaeological shows that the Wei bridge remains located out of the Chu gate of Chang'an city of Han dynasty.
This bridge is an important part of the city axis. It played an important role in the city traffic.

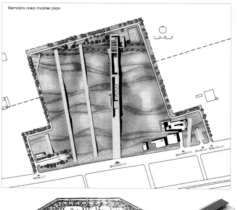

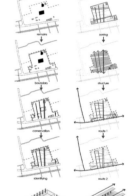

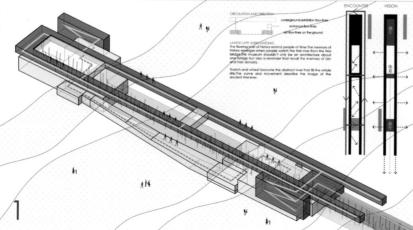

作品名称：中渭桥遗址博物馆设计
Museum Design of Wei Bridge remains of Qin and Han Dynasty

二等奖

院校名：西安建筑科技大学
设计人：王嘉琪，范岩，O. Faiyiga, A. Urbistondo
指导教师：苏静，常海青，同庆楠，吴涵儒，鲁旭，Albertus Wang, Martin Gold
课程名称：2015中美联合设计课程教学
作业完成日期：2015年06月
对外交流对象：美国佛罗里达大学建筑学院

An Architectural Exploration of the Axis and the Process of the RETROSPECT

中渭桥遗址博物馆设计

Museum design of Wei bridge remains of Qin and Han Dynasty

University: Xi'an University of Architecture and Technology
Dseigner: Wang Jiaqi, Fan Yan, O. Faiyiga, A. Urbistondo
Tutor: Su Jing, Chang Haiqin, Wu Hangru, Lu Xu, Albertus Wang, Martiin Gold
Course Name: Xi'an University of Architecture and Technology - University of Florida Joint Design Studio
Finished Time: Jun., 2015
Exchange institute: University of Florida School of Architecture

Second Prize

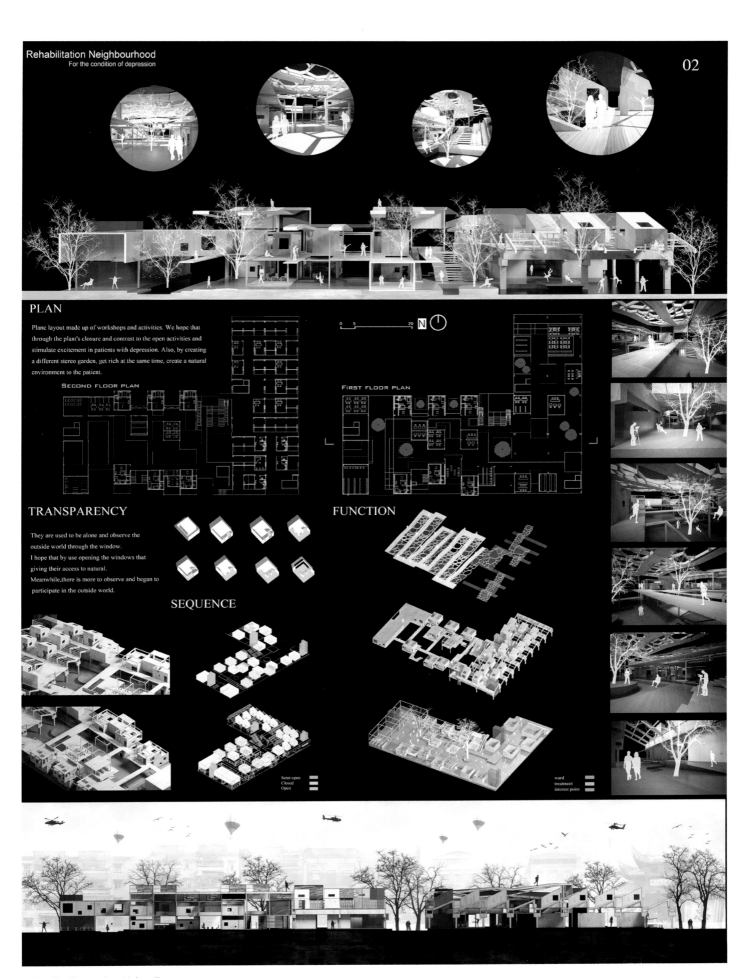

SPRINGSCAPE
SPRING THEME MUSEUM DESIGN 1

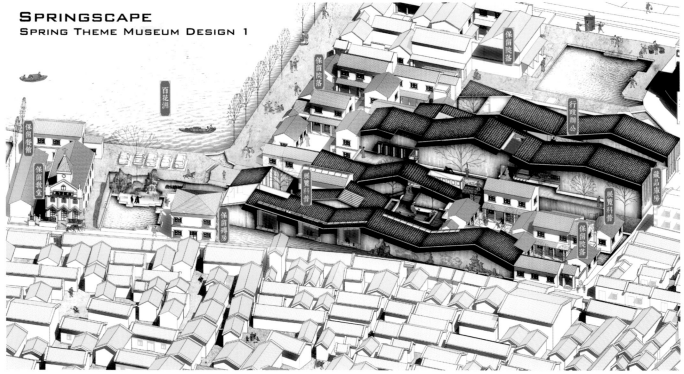

INTRODUCTION

THE SITE IS LOCATED IN JINAN TRADITIONAL HISTORIC AREA, THE PROJECT RESPECT THE HISTORIC FEATURES, THE SURROUNDING CITY TEXTURE AND THE EXISTING ENVIRONMENT--RETAIN THE NORTHWEST AND SOUTHEAST YARD, RESPECT THE LOCAL PEOPLE'S MEMORY; RETAIN THE WATER SYSTEM IN THE SITE, MAKE THE SPRINGS RETAINED BECOME A PART OF THE SPRING MUSEUM WHICH IS TO BE SHOWN, RETAIN THE STATUE IN THE SITE WHICH REFLECTS THE DAILY LIFE ABOUT THE SPRINGS, AS A PART OF THE EXHIBITION. BASED ON SURROUNDING ENVIRONMENT, THE PROJECT SELECTS THE FORM OF SLOPE ROOF BUILDING, BY INCREASING THE SLOPE ROOF FOLDED TO REDUCE THE MASS FOR CATERING TO THE SMAL-L-MASS BUILDING AROUND THE SITE.

LOCATION ANALYSIS

TYPE OF COMMERCE | FUNCTION DISTRIBUTION | BUILDING ENTRANCES | YARDS DISTRIBUTION

ACTUALITY ANALYSIS

SPRING IN THE SITE | TREES NEED TO BE RETAINED

STATUE RETAINED IN THE SITE

THE PROJECT NEED TO RESPECT HISTORY AND ORIGINAL ARENA, MAKE FULL USE OF THE RESETVED SPACE RESOURCES, RETAIN SPRING USED AS A PART OF THE EXHIBITION.

THE PROJECT NEED TO RESPECT HISTORY AND ORIGINAL ARENA, MAKE FULL USE OF THE RESETVED SPACE RESOURCES, RETAIN SPRING USED AS A PART OF THE EXHIBITION.

2003 | 2005 | 2010 | 2015

SINCE 2003, THE GOVERNMENT HAS DISMANTLED THE HOUSE, BULID NEW HOUSE, REPAIR OLD HOUSE IN THE SITE | 2003-2005, THE GOVERMENT HAS DISMANTLED A PART OF THE BUILDING IN THE SITE | 2003-2005, THE GOVERMENT HAS REPAIRED THE HOUSE WHICH HAS BEEN RETAINED | 2003-2005, FURTHER REPAIR THE HOUSE WHICH HAS BEEN RETAINED AND BUILD NEW YARDS ACCORDING TO THE TRADITIONAL LOCAL-STYLE DWELLING HOUSES OF JINAN

EXISTING BUILDING HIGHT ANALYSIS IN SITE

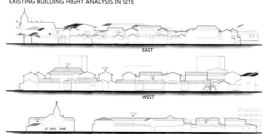

EAST
WEST
SOUTH

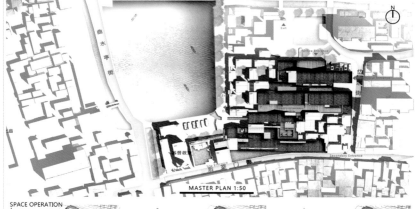

MASTER PLAN 1:50

SPACE OPERATION

1. ACTUALITY OF THE SITE
2. RETAIN THE SPRING AND SATATUE
3. RETAIN THE NORTHWEST AND SOUTHEAST YARD, RESPECT THE LOCAL PEOPLE'S MEMORY

4. ACCORDING TO RETAINED BUILDING, EXTRACT THE LINE OF CONTROL AND FILL THE SITE WITH LINEAR SPACE
5. ACCORDING TO RETAINED BUILDING, MAKE THE LINEAR SPACE SETBACK
6. BASED ON ACTUALITY OF THE SOUTH OF THE SITE THERE ARE SOME LOW LOCAL-STYLE DWELLING HOUSES, SO THE SOUTH MASS MIGHT BE LOW, ON THE CONTRARY, THE NORTH MASS MIGHT BE HIGH.

7. THE PROJECT SELECTS THE FORM OFSLOPE ROOF BUILDING, BY INCREASING THE SLOPE ROOF FOLDED TO REDUCE THE MASS FOR CATERINGTO THE SMAL L-MASS BUILDINGS AROUND THE SITE.
8. MAKE THE PART OF THE ELEVATION ALONG THE HOUZAIMEN STREET SETBACK, ADD THE EXHIBITION FUNCTION, LET THE EXHIBITION FUNCTION EXTEND OUTSIDE
9. DEEPEN THE SITE AND PERFECT THE DETAILS

1-1 SECTION 1:500

SOUTH ELEVATION 1:500

作品名称：泉景——泉水博物馆设计
Springscape—Spring Theme Museum Design

二等奖

院校名：山东建筑大学
设计人：王天元
指导教师：房文博，孔亚暐，Tony Van Raat
课程名称：建筑设计4
作业完成日期：2015年06月
对外交流对象：新西兰UNITEC理工学院

SPRINGSCAPE
SPRING THEME MUSEUM DESIGN 2

PERSPECTIVE SECTION 2-2

MASS OPERATION

Actuality: There are some small-mass houses which are harmonized with the surrounding houses in the site.

How can both meet the requirements of museum of large space, and fit the small mass buildings around skin texture form is where the plan needs to be thinking about.

Through increasing the slope roof folded across several to reduce cubic construction. The project can cater to the surrounding texture.

ELEVATION OPERATION

How to deal with the old trees retained needs to be thinking about in the project.

leave the elevation blank, without any decoration in the elevation, make the elevation to be the background to set off the old trees.

The shadow which is changed with time on the elevation can be a part of exhibition. Enrich the exhibition in the museum.

MATERIAL OPERATION

How can the springs in the site to be dealed with needs to be thinking about in the project

The project through material selection to take advantage of retained springs: concrete which represents thick with a strong contrast between springs which represents light.

The springs contrasted with concrete which represents thick stands out in the historical area.

AXONOMETRIC SECTION ANALYSIS

Interior exhibition 1

Interior exhibition 2

Interior exhibition 3

AXONOMETRIC ANALYSIS

Interior exhibition hall

Outdoor exhibition

Roof platform

INTERIOR PERSPECTIVE

INTERIOR PERSPECTIVE **OUTDOOR PERSPECTIVE**

AXONOMETRIC EXPLOSION DIAGRAM

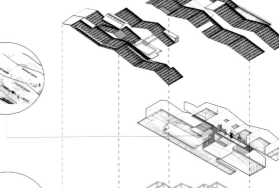

1 The church
2 Souvenir shop
3 Duty room
4 Resting place
5 Service counter
6 The hall
7 Temporary exhibition
8 Exhibition near the water
9 Fixed exhibition
10 Temporary exhibition
11 Exhibition near the water
12 Small multimedia hall
13 Spring culture theme exhibitions
14 Jinan old city model
15 Traditional customs exhibition
16 The statue retained
17 Warehouse
18 The laboratory
19 Reference room
20 Duty room
21 Photographic studio
22 Print room
23 Outdoor activities
24 Tea house
25 The bookstore
26 Celebrity exhibit
27 Rest platform
28 Overhead
29 Fixed exhibition
30 Above the yard
31 Office
32 Meeting romm
33 Experts studio
34 Director's Office

F2 PLAN 1:300

F1 PLAN 1:300

University: Shandong Jianzhu University
Designer: Wang Tianyuan
Tutor: Fang Wenbo, Kong Yawei, Tony Van Raat
Course Name: Spring Theme Museum Design
Finished Time: Jun., 2015
Exchange Institute: New Zealand UNITEC Institute of Technology

Second Prize

BOGOTA IN DUALITY
A Housing-Landscape Complex in-between the Old and New, Bogota, Colombia

01

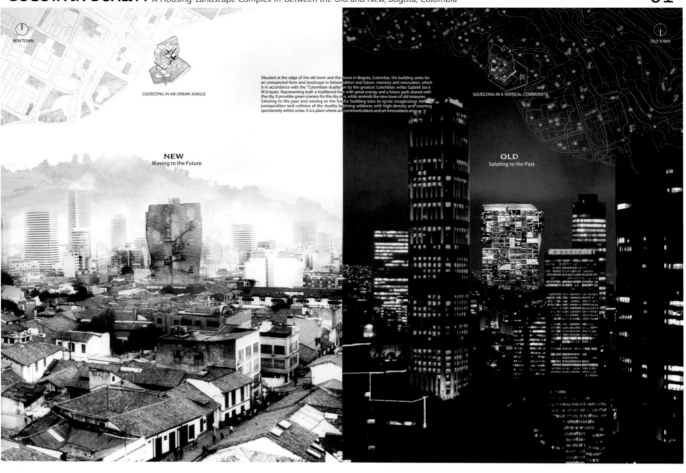

SQUEEZING IN DUALITY

The building in duality provides a link between a wild green park and a crowded vivid city. From one side, a vertical grid made up of "districts", "squares" and "landmarks" reminds the gentrified people in the new town of the memories of old heritage. While from the other side, a thick green wall with vertical linking freeways and entertaining facilities symbolizes the jungle of Colombia, making it a nice park for the dense old city.

SQUEEZING IN-BETWEEN

In-between the "city" and the "forest" lies the transitional space, which contains necessary public and traffic functions, such as the entrance lobby, with more accessibility to both parts of the building. A delicate structural system is designed to integrate the vertical green wall with the apartment building. Hanging stairs are inserted between different courtyards and public space.

Standing at the edge of the new and the old city, the building is designed as a merge of spatial and functional characteristics of both sides, trying to stimulate communication and chemistry between social classes. The model itself, using recycled daily materials, also resonates with this orientation.

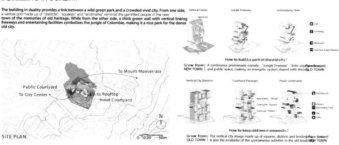

作品名称：双面的波哥大
Bogota in Duality

院校名称：清华大学
设计人：熊哲昆
指导教师：程晓青，邹欢
课程名称：AIAC六国九校国际学生建筑设计工作坊
作业完成日期：2014年10月
对外交流对象：哥伦比亚安第斯大学，法国拉维莱特大学，意大利威尼斯建筑大学等

二等奖

28

BOGOTA IN DUALITY
A Housing-Landscape Complex in-between the Old and New, Bogota, Colombia

02

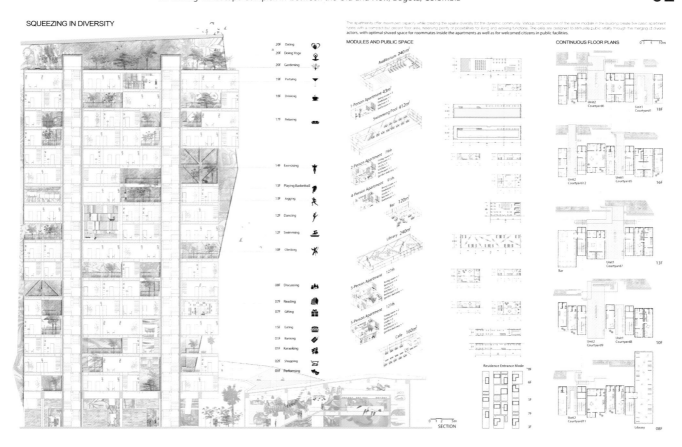

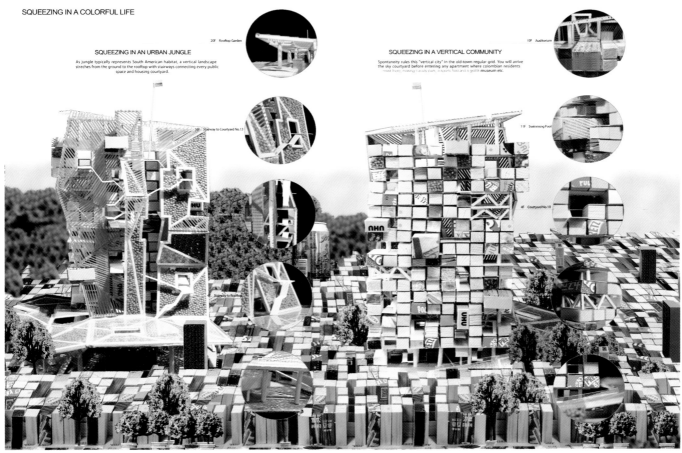

University: Tsinghua University
Designer: Xiong Zhekun
Tutor: Cheng Xiaoqing, Zou Huan
Course Name: AIAC International Architecture Design Studio
Finished Time: Oct., 2014
Exchange Institute: Universidad de los Andes, and so on

Second Prize

LEUCHT TURM

executed in Aug. 2015

"Open before us is the architects' plan. Next to it is a sheet of music. The architect fleetingly reads their composition as a structure of elements and spaces in their light."

----Louis Kahn

01

INTRODUCTION

In our exchange program in Hochshule Luzern in Switzerland, we were so lucky to find that our project was a realistic one, a "Light Tower" that should become a symbol of the local cultural project "Halbinsel" in Horw for one year. The site is small, but there must be kind of an unusual thing that attract people's eyes and tells them, this is not a normal building in common sense, but a landmark. Client required that basically the building would provide spaces for exhibition, lectures and information of current events in Horw. The building must be expressive enough while the budget was very limited. Prof. Dieter Geissbühler was leading the 11-people class to do the project. The semester started in February. Firstly we worked for 3 weeks individually to have 11 ideas, then the teachers would choose 3-5 ideas to continue, and students would team up and work together. After another 2 month the teachers and clients would discuss and decide which one to be executed. The small tower was planned to be standing there in the end of August 2015. Besides Prof. Geissbühler, we had Yves Dusseiller, Tina Unruh, Prof. Hansjürg Buchmeier, Uwe Teutsch as our instructors.

SITE

The site locates in a parking lot near the bus station "Horw Zentrum", just a minute walk towards Migros, Coop, Kantanal Bank and Reiffessen. Also, the local church is just about a hundred meters away. In deed this is the center of Horw and plenty of people walk pass by everyday. Due to the finance and the regulations, we were informed that the building must not be higher than 20 meters. Another settled requirement was, the fundamental material must be wood, and it is better to be executed by the traditional wood construction method. Also, we were suggested to introduce natural wood in the project. Thinking of its functions and temporality, We started to try to build a fascinating open structure that seems to absorb things in.

- Shear Walls as Exhibition Boards
- Natural Wood Scaffolding
- Core Platform for walking and looking

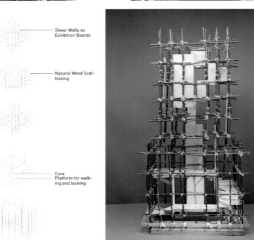

Traditional Chinese scaffolding is made by long bamboos. They are simple, easy to be made and quite stable. One important feature is that the bamboo pieces are connected by rope knots. Skillful workers can make a knot very efficiently. Thus the bamboo scaffolding saves time and money for the investigators.

CONCEPT

We both agreed that the building should create an atmosphere of the unfinished and continued the scaffolding concept. During the critic I acknowledged the 3-layers scaffolding was too complicated and expensive. Additionally, People should not reach higher than 3 meters because of fire regulation in the situation of this project. The design was reduced and simplified to pure scaffolding. As well as the material, changed from smooth artificial wood to natural wood trunks that we could collect in a nearby forest, connected by rope knot just like the way Chinese do. After a series of adjustment, our design became the only one that is not only expressive and fit the demands of the clients, but also cheap enough to be built within low budget.

After being selected as the final project to be executed, we collaborated with the professors, carpenters, engineers and materials suppliers etc. Information came from multiple parties to one project which needed to be adapted and fit the requirement. The final solution is even more reduced while the concept still lies within.

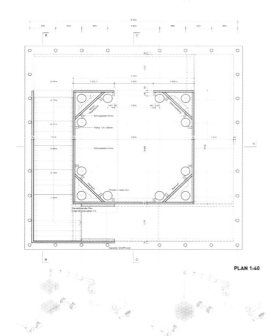

PLAN 1:40

CORNER DETAILS　　STEP DETAILS

EVOLUTION OF THE ELEVATION

FORMER CUNSTRUCTION PROGRESS

作品名称：灯塔　Leucht Turm

院校名：中央美术学院
设计人：李师嵘，王丰
指导教师：Dieter Geissbühler, Yves Dusseiller, Tina Unruh, Hansjürg Buchmeier, Uwe Teutsch
课程名称：临时展览木塔设计
作业完成日期：2015年08月
对外交流对象：瑞士卢塞恩应用科学与艺术大学建筑系

二等奖

LEUCHT TURM

Details design is extremely important especially in Switzerland, where everything is quite accurate comparing to China. Mainly under the instruction of the experienced carpenter, architect, Yves Dusseiller, we carefully made drawings of the stairs, the base, the corners etc. The factors of rain proof, subtle height difference in one step, how the trunks are fixed on the wooden board are pretty well considered.

The construction were planned to start at the end of June 2015 and finish at the end of August. We couldn't participate in the wood construction part because of the insurance matters, but we joined the carpenter apprentices tin the forest to take the skin off the trunks. We were also there to attend the meeting with the wood supplier. Always being in the progress of a project from the beginning to the end makes an incredible experience.

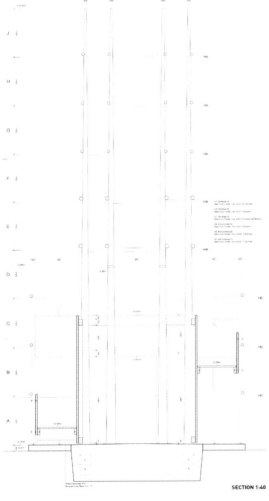

We had a long discussion through months with the engineers about how to make the tower stiff. One always propose to use steel ropes diagonally installed in between the trunks. But we didn't like it for it disturbed the expression of a pure scaffolding. Insisting and looking for solutions, we were finally able to persuade the engineer that 12 vertical pieces of trunks that continued from the base to the top, fixed by artificial walls in the ground floor could make the tower stiff enough.

The stairs were no longer designed in the corner but in the place where there was supposed to be the platform, for this could be a lot easier to be produced and construct on site. Each step is approximately 80cm wide and allows people to slow down. The sky is opened up from the ground floor where natural light creates nice atmosphere in the octagonal room. Now we have only one small platform on the second floor for lectures.

We used to have a small concrete base for every trunk but now we make it a whole. The 12 trunks in the center are inserted from the underground and fixed by concrete, while the other short pieces are supported by small metal from the ground to avoid the corrosion of rain water. On account of there is no Chinese scaffolding technique in Europe, rope knots are replaced by metal joints to ensure safety.

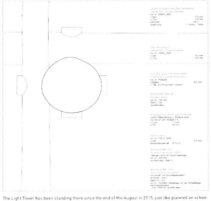

SECTION 1:40

The Light Tower has been standing there since the end of the August in 2015, just like planned on schedule. And we were so glad that we could see it built before we returned to China. This is the first time that we participate in one project from an idea to reality. Half a year we started to know how difficult and complicated if a project wanted to come to the earth, even a small simple one. And we realize that the role an architect plays is not only designer, but the person who must communicate with clients, builders and suppliers etc. He or she must balance all of the requirements, and meanwhile, keep the project from deviating its original concept.

FINAL CUNSTRUCTION PROGRESS

University: China central Academy of fine Arts
Designer: LI Shiyao, WANG Feng
Tutor: Dieter Geissbühler, Yves Dusseiller, Tina Unruh, Hansjürg Buchmeier, Uwe Teutsch
Course Name: Focus on Material - Hochschule Luzern - Technik & Architekturco. Ltd. Joint Design Studio
Finished Time: Aug., 2015
Exchange Institute: Lucerne University of Applied Sciences and Arts

Second Prize

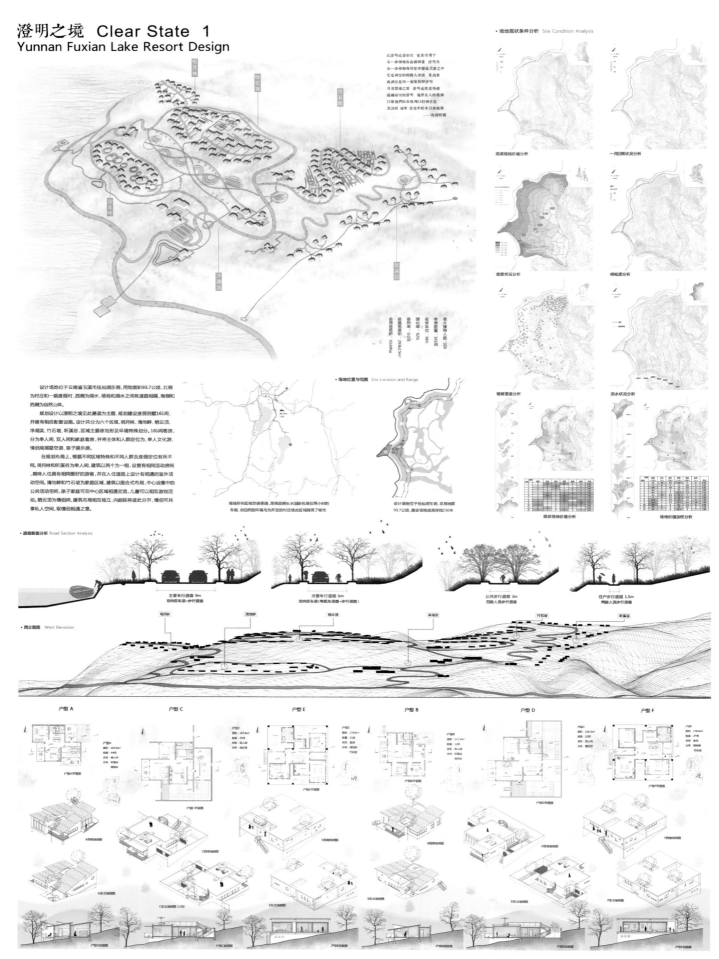

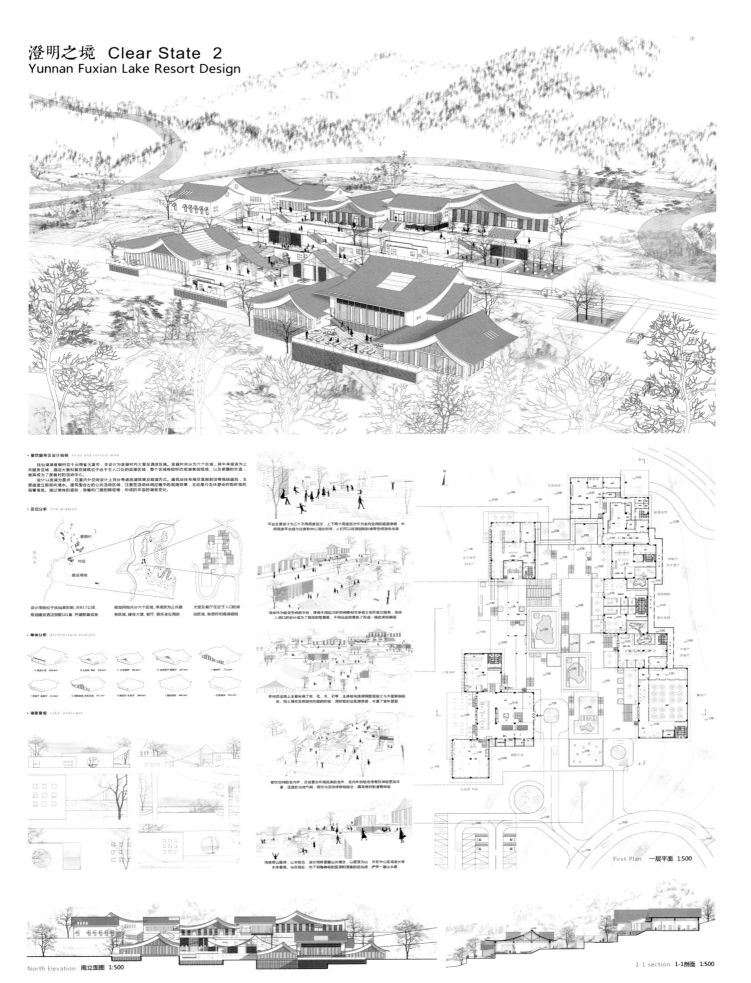

Back to the future of Tang Dynasty (A)
Tang Axis the Small Wild Goose Pagoda Area Urban Design

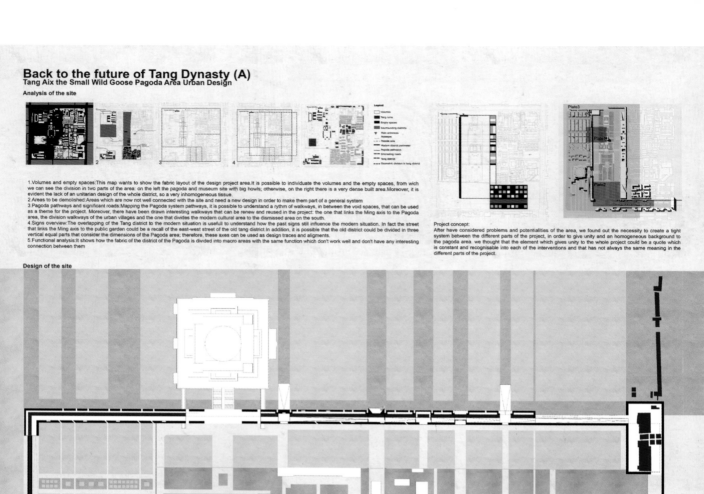

Analysis of the site

1. **Volumes and empty spaces:** This map wants to show the fabric layout of the design project area. It is possible to individuate the volumes and the empty spaces, from wich we can see the division in two parts of the area: on the left the pagoda and museum site with big howls; otherwise, on the right there is a very dense built area. Moreover, it is evident the lack of an unitarian design of the whole district, so a very inhomogeneous tissue.
2. **Areas to be demolished:** Areas which are now not well connected with the site and need a new design in order to make them part of a general system.
3. **Pagoda pathways and significant roads:** Mapping the Pagoda system pathways, it is possible to understand a rythm of walkways, in between the void spaces, that can be used as a theme for the project. Moreover, there have been drawn interesting walkways that can be renew and reused in the project: the one that links the Ming axis to the Pagoda area, the division walkways of the urban villages and the one that divides the modern cultural area to the dismissed area on the south.
4. **Signs overview:** The overlapping of the Tang district to the modern situation is usefull to understand how the past signs still influence the modern situation. In fact the street that links the Ming axis to the public garden could be a recall of the east-west street of the old tang district. In addition, it is possible that the old district could be divided in three vertical equal parts that consider the dimensions of the Pagoda area; therefore, these axes can be used as design traces and aligments.
5. **Functional analysis:** It shows how the fabric of the district of the Pagoda is divided into macro areas with the same function which don't work well and don't have any interesting connection between them

Project concept:
After have considered problems and potentialities of the area, we found out the necessity to create a tight system between the different parts of the project, in order to give unity and an homogeneous background to the pagoda area. we thought that the element which gives unity to the whole project could be a quote which is constant and recognisable into each of the interventions and that has not always the same meaning in the different parts of the project.

Design of the site

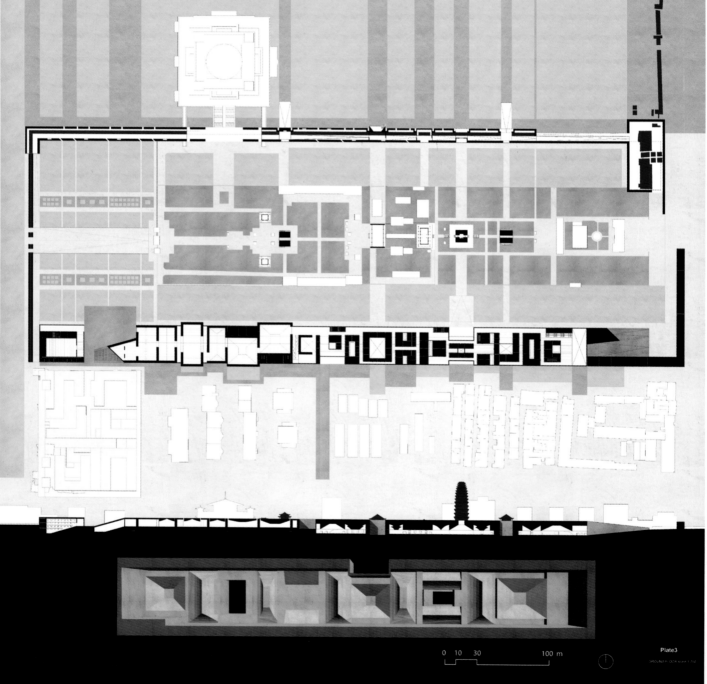

作品名称：回溯唐的未来——唐轴线小雁塔地段城市设计
Back to the Future of Tang Dynasty

院校名：西安建筑科技大学
设计人：侯帅，Giulia Mazzuchelli, Costanza Mondani, Miriam Pozolli, 赵晋, 赵彬彬
指导教师：李昊，常海青，鲁旭，李焜，laura Anna Pezzetti, Carlo Palazzolov
课程名称：2015西安建筑科技大学—意大利米兰理工大学国际联合工作营课程教学
作业完成日期：2015年05月
对外交流对象：意大利米兰理工大学建筑学院

二等奖

Back to the future of Tang Dynasty (B)
Tang Aix the Small Wild Goose Pagoda Area Urban Design

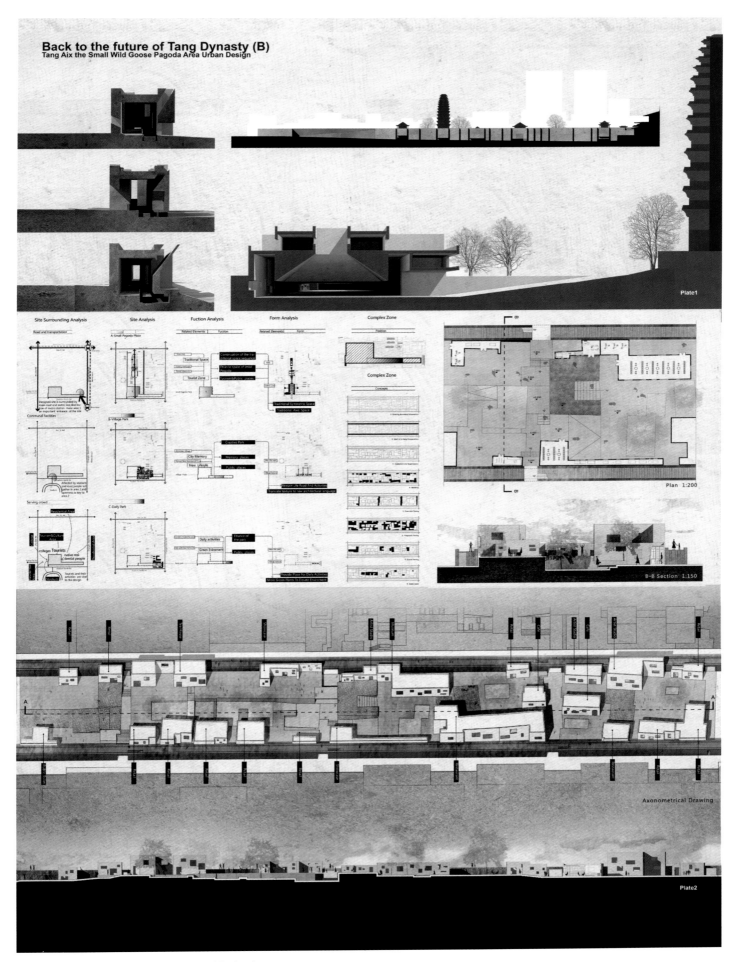

University: Xi'an University of Architecture and Technology
Designer: Hou Shuai Giulia Mazzucchelli, Costanza Mondani, Miriam Pozzoli, Hou Shuai, Zhao Jin, Zhao Binbin
Tutor: Li Hao, Chang Haiqing, Lu Xu, Li Kun, Laura Ana Pezzetti, Carlo Palazzolo
Course Name: 2015 Xi'an University of Architecture and Technology - Politecnico di Milano Interenational Workshop
Finished Time: May, 2015
Exchange Institute: Politecnico di Milano

Second Prize

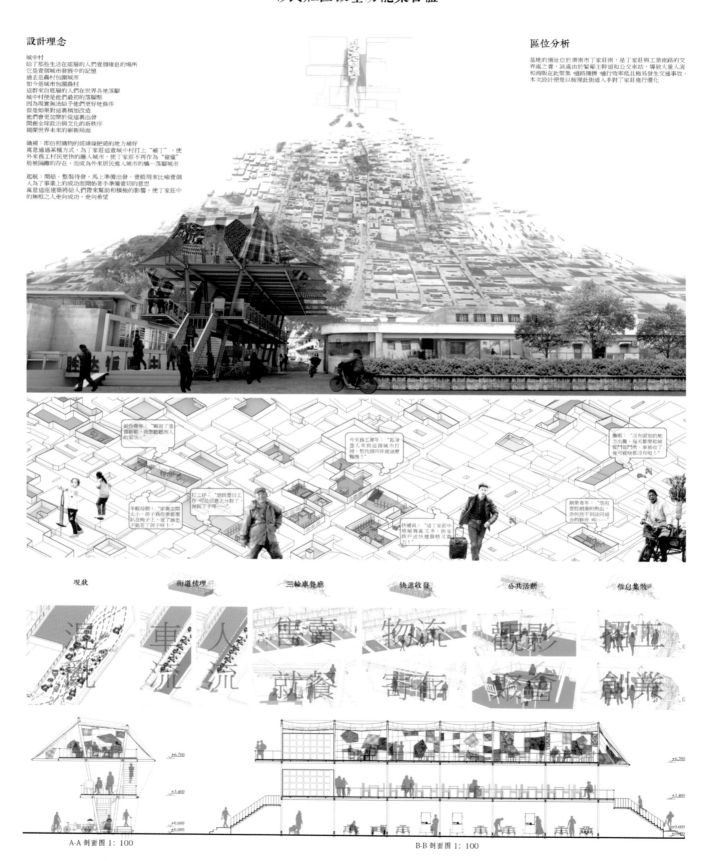

作品名称：织补·起航——移民社区微型功能集合体
Darning Set Sail

院校名：山东建筑大学
设计人：贾鹏
指导教师：金文妍，张雅丽，Tony Burge
课程名称：建筑设计4
作业完成日期：2014年10月
对外交流对象：新西兰UNITEC理工学院

二等奖

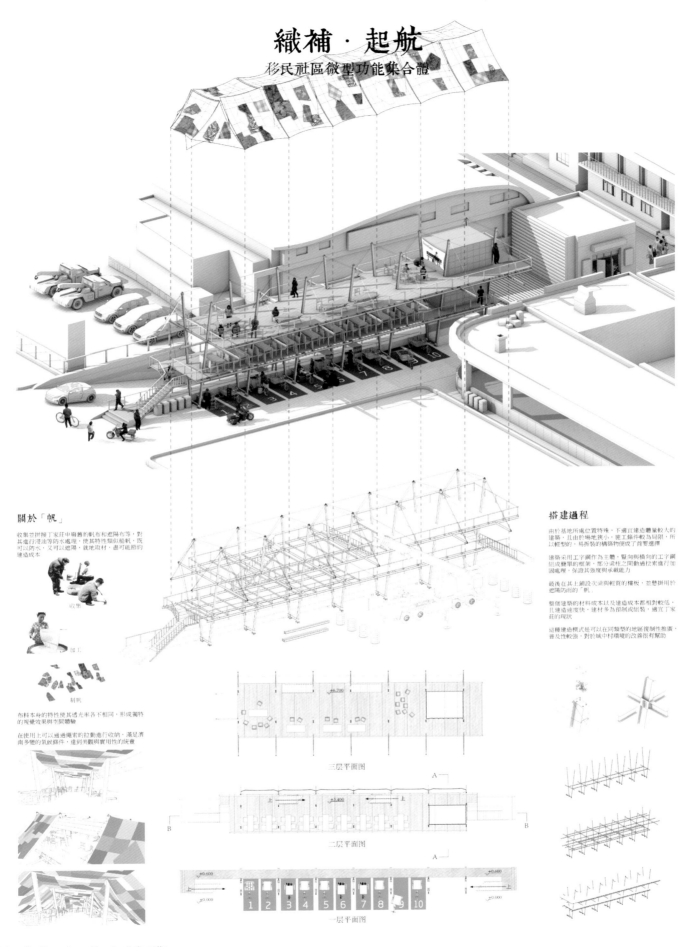

University: Shandong Jianzhu University
Designer: Jia Peng
Tutor: Jin Wenyan, Zhang Yali, Tony Burge
Course Name: Shandong Jianzhu University - New Zealand UNITEC Institute of Technology Joint Design Studio
Finished Time: Oct., 2014
Exchange Institute: New Zealand UNITEC Institute of Technology

Second Prize

Regional Architecuture –
Industrial reconstraction and renewal design of changsha papermaking factory

the Site Analysis:
The site is located in the west bank of xiangjiang river in changsha of China's hunan province. The Tianlun papermaking factory belongs to the industrial cultural heritage of old China. it distributed the abandoned workshop buildings, which are different areas and rich in forms and space.

Concept Source:
The inspiration come from the remaining enclosed and partly enclosed yard, patio, veranda space of the workshop buildings. Then I thought about the courtyard and "patio(tian jin)" of the Chinese south regional architecture, and connected with the regional climate characteristics to make a industrial reconstraction and renew design.

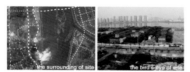

Form Generation: **reconstruction in the workshop building:**   **Structure Analysis:**

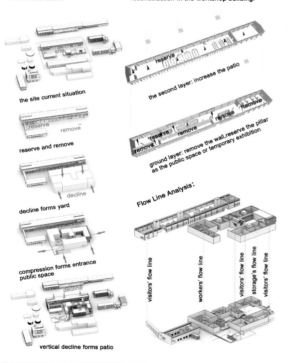

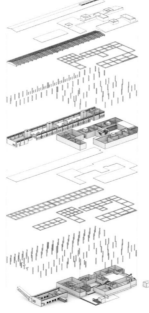

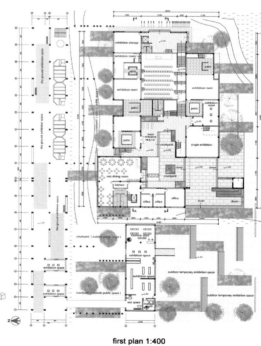

first plan 1:400

section1-1 1:200

作品名称：长沙滨江天伦造纸厂更新改造联合设计
Regional Architecture—Industrial Reconstraction and Renewal Design of Changsha Papermaking Factory

院校名：湖南大学
设计人：何磊
指导教师：罗荩，陈翚，李旭
课程名称：建筑设计A4
作业完成日期：2015年07月
对外交流对象：捷克共和国捷克技术大学建筑学院

二等奖

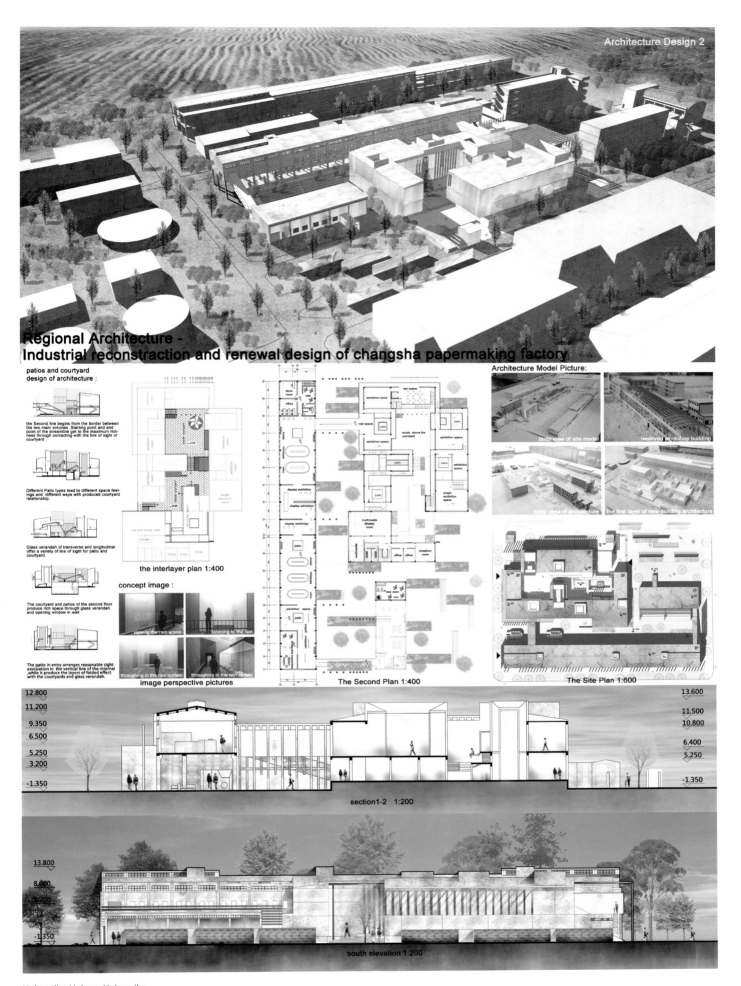

University: Hu'nan University
Designer: He Lei
Tutor: Luo Jing, Chen Hui, Li Xu
Course Name: Hunan - Czech Technical University Joint Design Studio
Finished Time: Jul., 2015
Exchange Institute: Czech Technical University School of Architecture

Second Prize

南京大行宫碑亭巷旧居住区更新改造设计
THE RENOVATION OF NANJING BEITING LANE RESIDENTIAL DISTRICT

场地调研 SITE INVESTIGATION 01

联合设计课程目标
通过研究中国当下中国高速发展的城市中的典型街区的设计更新方法，从而理解如何进行正确的设计，而不仅仅是进行外观的美化。这需要建筑师们站在使用者的角度体会建筑与功能的关系。正确的设计更要求建筑师们对城市和建筑的设计具有一种策略性的导向，而不仅是灵光一现。城市设计的作用已被广泛接受，被认为是一种可指导某一区域的成功发展并能惠及区域内每一个个体的有效策略。在当代中国，城市高速发展，且涉及社会多个层面问题，传统的把建筑师看作英雄式角色的观点完全不适用了，而城市设计师式的更加从整体层面、策略层面去思考城市空间发展的思路显得更加明智。

基地环境
基地处于南京城市中心区域，基地内部为上世纪80年代建造的居住小区，周边地块则是城市快速发展后新建的重要城市建筑和设施，如1912休闲街区、南京图书馆、地铁2号3号线等。

本方案设计思路
通过空间的整改和遗存的保护为活动的发生提供场所，同时让艺术活动植入居民日常，依据不同的艺术文化活动所需的空间形式，开展空间与活动功能的设计。

Target of the studio
The study of the typical Chinese blocks which are under the rapid developing period. It calls for a correct design, not only an optical shining design. If we take the chance to stand with the client in thinking about architecture, the demand on correctness requires the architects to start from design strategies, rather than sparks of idea, and that base firms the foundation for the development of a project. The role of urban design, which is commonly understood and accepted. A clear framework of planning strategy could guide a district to success with the participation of different developing individuals. Architects are no more lonely heroes. Seeking strategic resolution on design methods as a urban designer from a general pespective for the urban development is more sensible.

Site condition
The site chosen for the studio is located in the center of Nanjing. There was a long story about the history in this area, but now it's filled with the residential buildings built around 1980s. Also, there are many important historical and cultural buildings around here, for example, the 1912 commercial and creation area, the President Palace, the Nanjing library, Jiangning Weaving Museum etc.

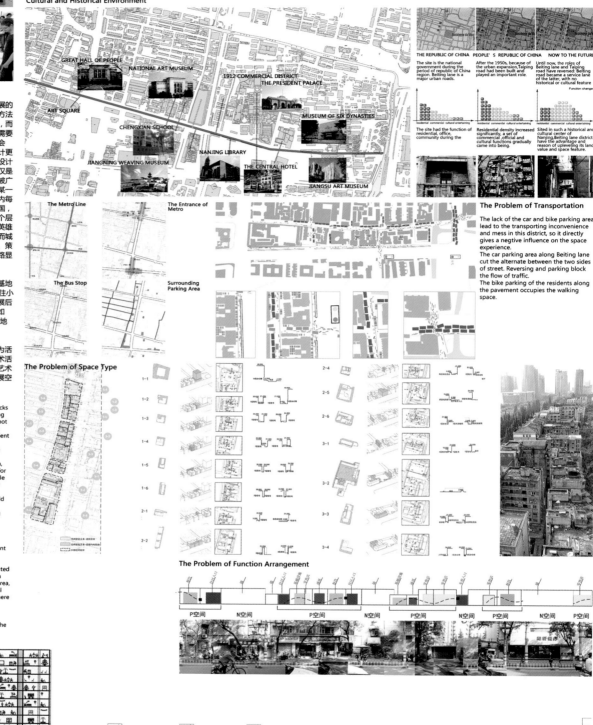

作品名称：南京大行宫碑亭巷旧居住区更新改造设计
The Renovation of Nanjing Beiting Lane Residential District

二等奖

院校名：东南大学
设计人：宋文颖，白宇泓，黄迪奇，Victoria Einrach
指导教师：韩晓峰，葛明
课程名称：建筑设计
作业完成日期：2014年11月
对外交流对象：奥地利维也纳工业大学

南京大行宫碑亭巷旧居住区更新改造设计
THE RENOVATION OF NANJING BEITING LANE RESIDENTIAL DISTRICT

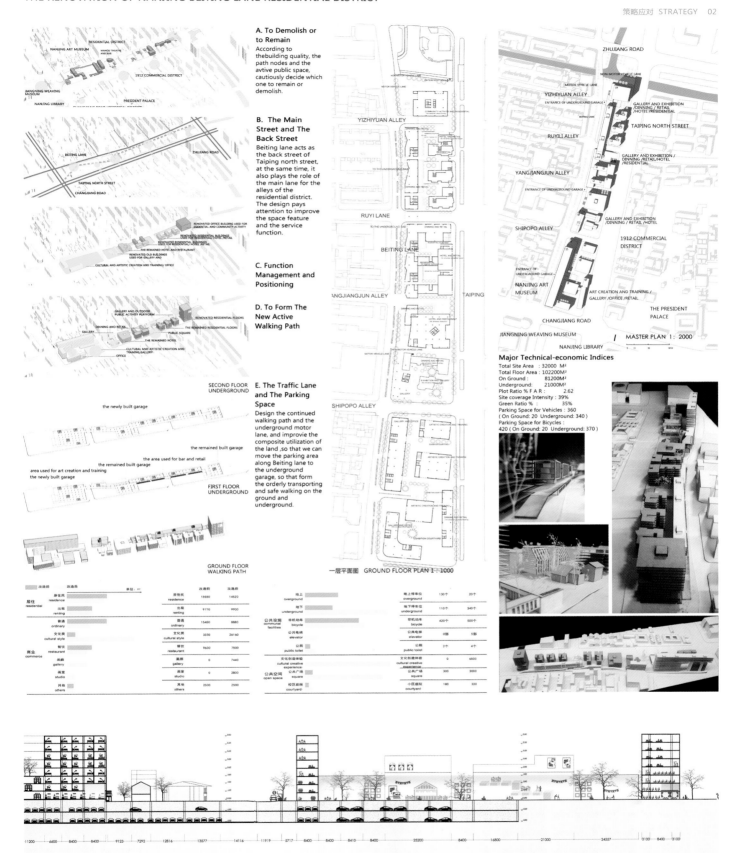

University: Southeast University
Designer: Song Wenying, Bai Yuhong, Huang Diqi, Victria Einrach
Tutor: Han Xiaofeng, Ge Ming
Course Name: SEU-TU Wien - Tongji - Shenzhen University Joint Study Studio
Finished Time: Nov., 2014
Exchange Institute: Technische Universitat Wien

Second Prize

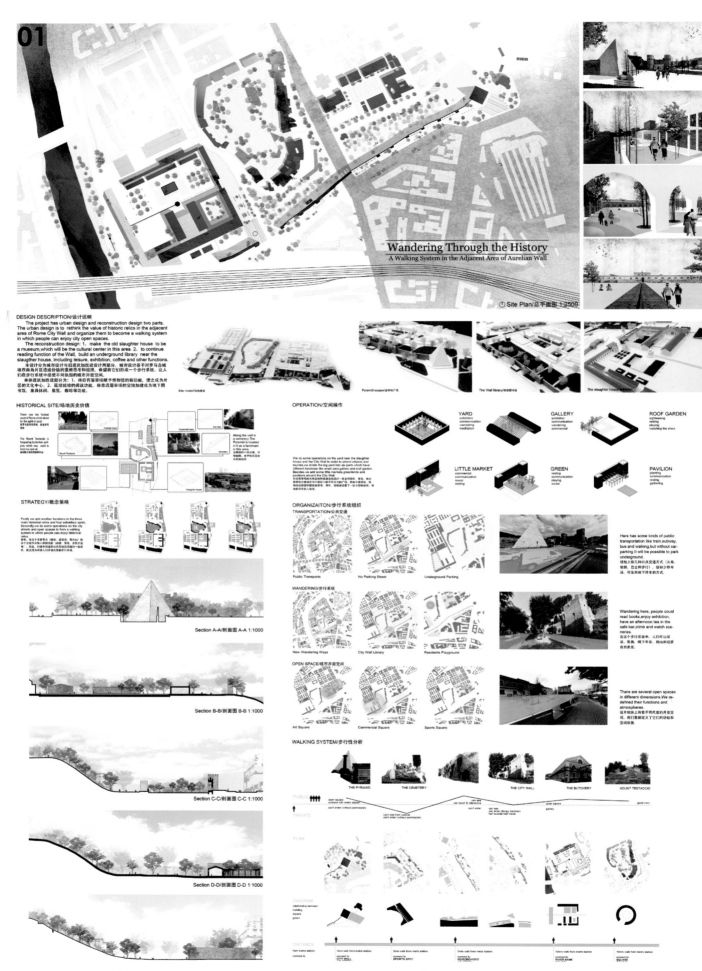

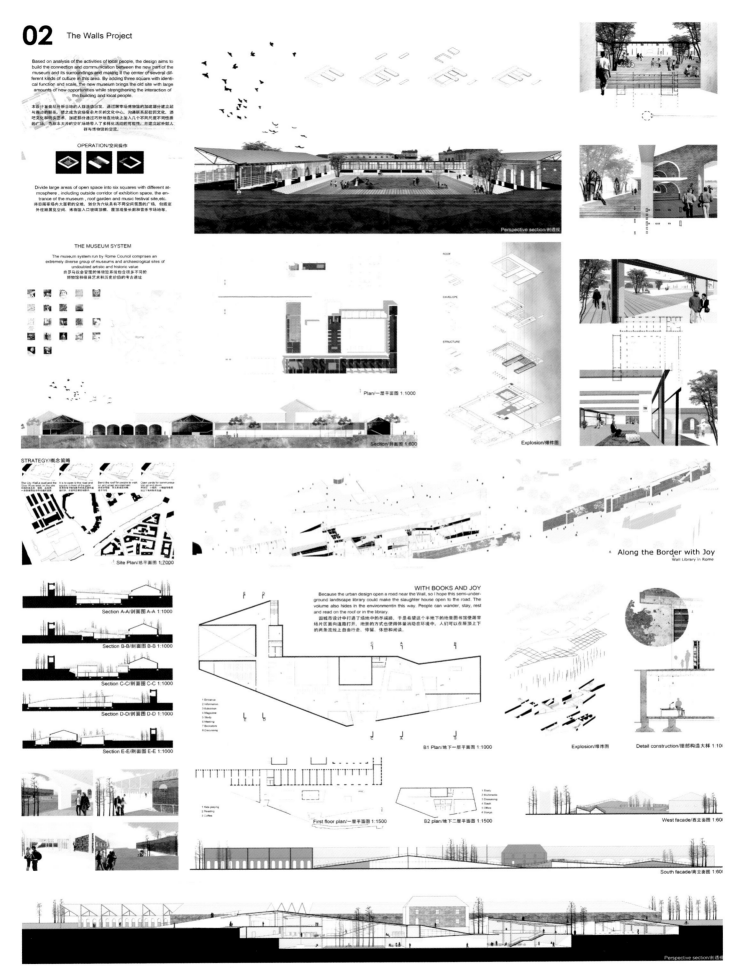

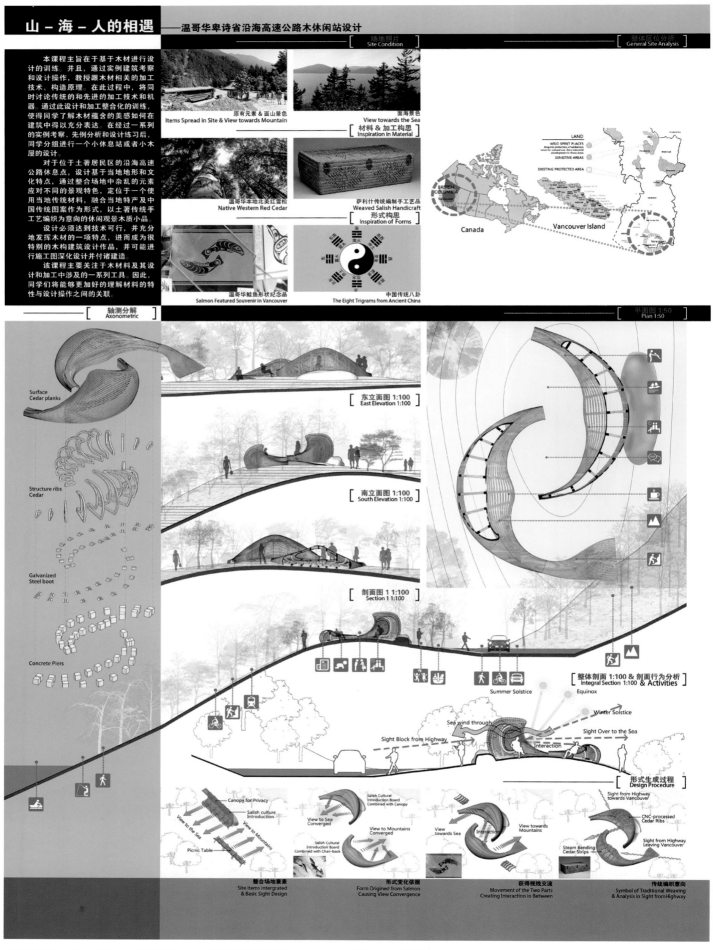

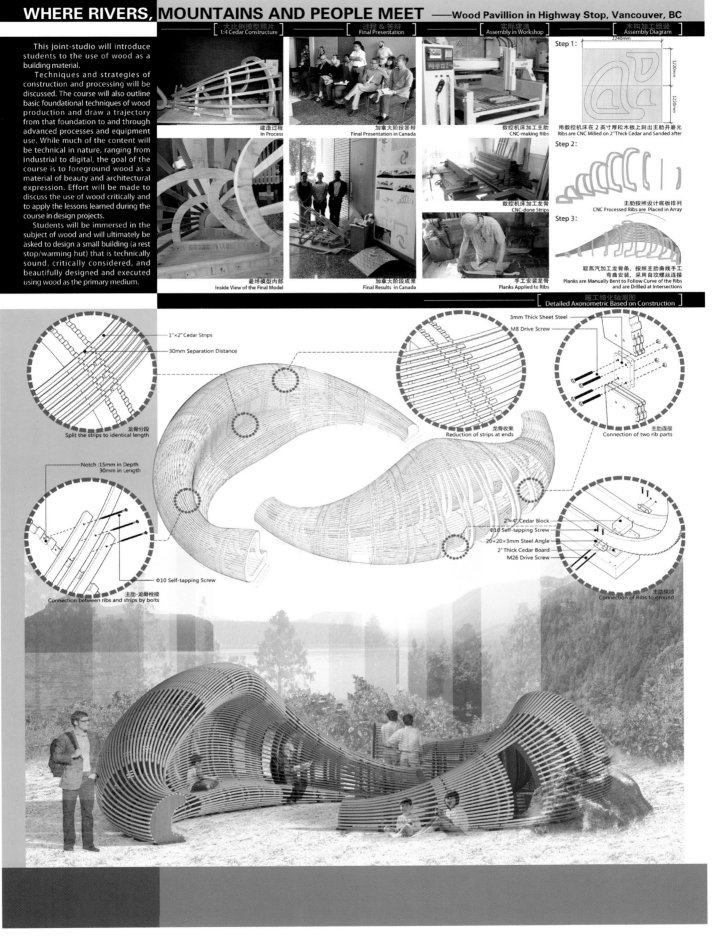

WHERE RIVERS, MOUNTAINS AND PEOPLE MEET —— Wood Pavillion in Highway Stop, Vancouver, BC

This joint-studio will introduce students to the use of wood as a building material.

Techniques and strategies of construction and processing will be discussed. The course will also outline basic foundational techniques of wood production and draw a trajectory from that foundation to and through advanced processes and equipment use. While much of the content will be technical in nature, ranging from industrial to digital, the goal of the course is to foreground wood as a material of beauty and architectural expression. Effort will be made to discuss the use of wood critically and to apply the lessons learned during the course in design projects.

Students will be immersed in the subject of wood and will ultimately be asked to design a small building (a rest stop/warming hut) that is technically sound, critically considered, and beautifully designed and executed using wood as the primary medium.

University: Southeast University
Designer: Zong Yuanyue, Ramona Montecillo, Robert Maggay
Tutor: Han Xiaofeng, Tu Susan
Course Name: SEU - UBC - Canada Wood Joint Design and Constraction Studio
Finished Time: Aug., 2015
Exchange Institute: UBC School of Architecture, Canada Wood

Second Prize

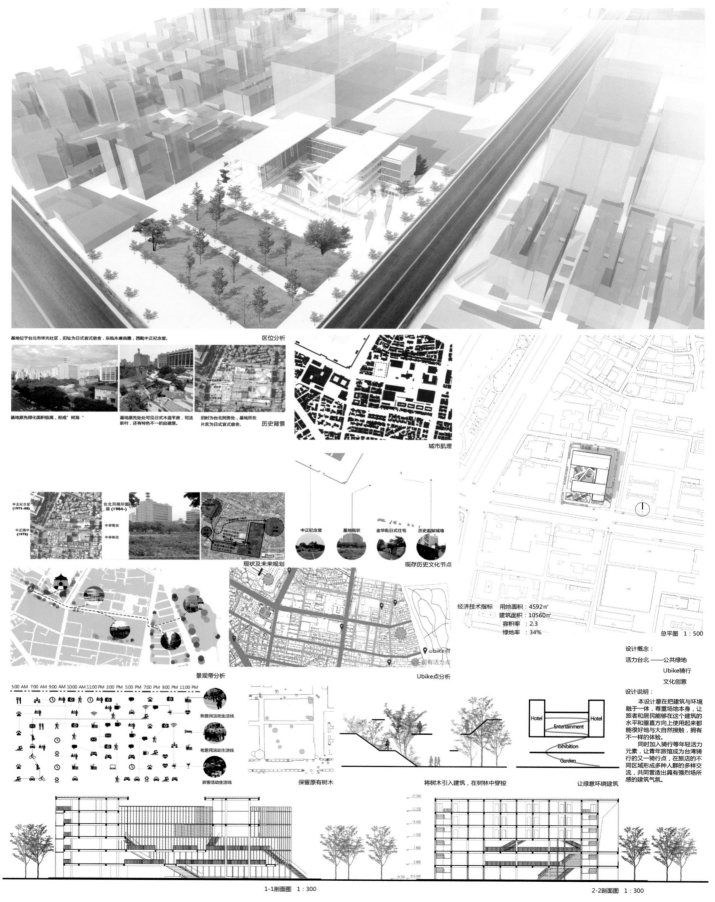

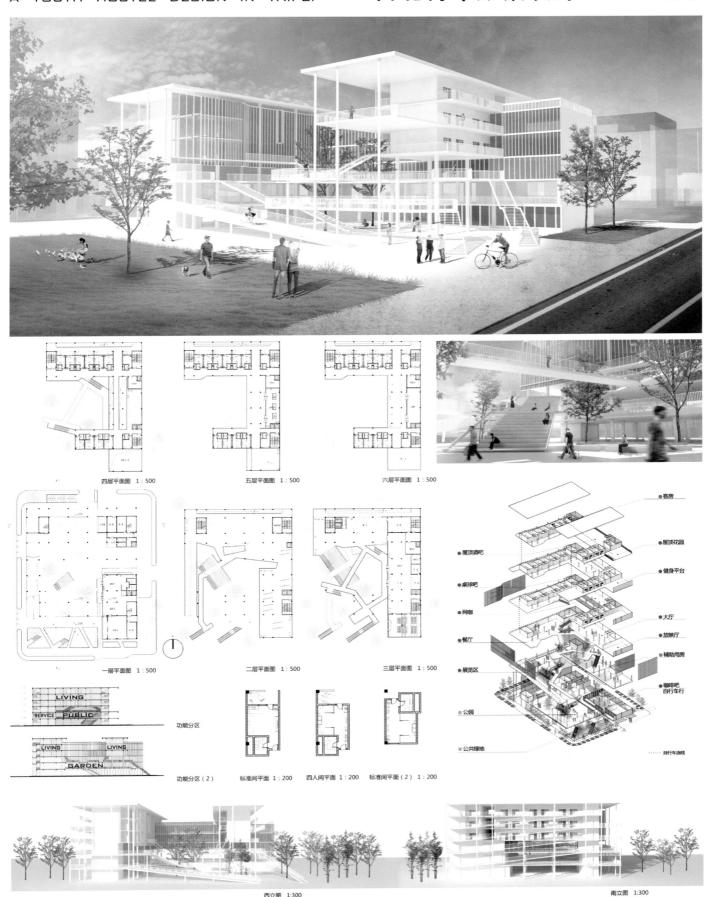

Floating Blocks
the New Riverside District
LIBRARY AND INFORMATION CENTER OF THE COMMUNITY

01

The base is a disused factory which is located in the New Riverside District of Changsha. This area used to be a industrial area in the late 19th century, and now, it is supposed to be a central business district with ecological node and living area.

1-1section

2-2section

Site Plan

Design Concept

As a library&information center of the community based on a abandoned factory, this project focuses on the inheritance of old individual culture. Another important goal of the design is to bring the internal program of this building to the environment and nearby residents. Breaking the traditional institutional image of the library as a closed building, it will be a place which participates in the urban daily life of this area with openness and diversification.

functions

concept

the new one (as library) + the old one (for exhibition)

break & set into / creat bottom overhead space for landscape & activity place

plan

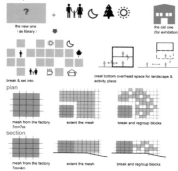

mesh from the factory 7m×7m / extent the mesh / break and regroup blocks

section

mesh from the factory 7m×4m / extent the mesh / break and regroup blocks

reconstruction of the factory

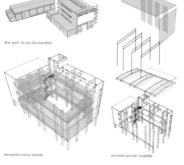

the part to be dismantled

reconstruction inside / reconstruction outside

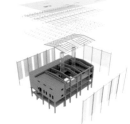
Reconstruction of The Factory

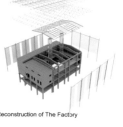
The Construction of The Wall

The Construction of Pavilion

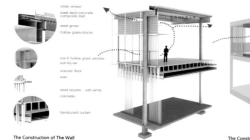

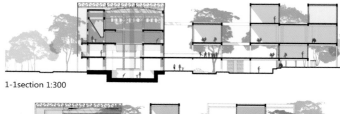
1-1section 1:300

2-2section 1:300

作品名称：悬浮森林——滨江新城厂房改扩建之图文媒体中心
Floating Blocks-Library and Information Centor of the community

院校名：湖南大学
设计人：廖若微
指导教师：严湘琦，张蔚
课程名称：建筑设计A4
作业完成日期：2015年01月
对外交流对象：捷克共和国捷克技术大学建筑学院

二等奖

Floating Blocks
LIBRARY AND INFORMATION CENTER OF THE COMMUNITY

02

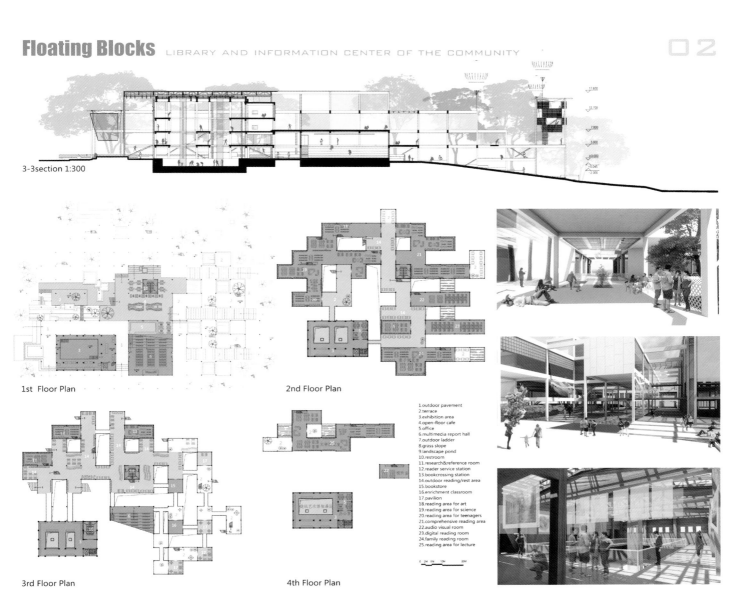

3-3 section 1:300

1st Floor Plan
2nd Floor Plan
3rd Floor Plan
4th Floor Plan

1. outdoor pavement
2. terrace
3. exhibition area
4. open-floor cafe
5. office
6. multimedia report hall
7. outdoor ladder
8. grass slope
9. landscape pond
10. restroom
11. research&reference room
12. reader service station
13. bookcrossing station
14. outdoor reading/rest area
15. bookstore
16. enrichment classroom
17. pavilion
18. reading area for art
19. reading area for science
20. reading area for teenagers
21. comprehensive reading area
22. audio visual room
23. digital reading room
24. family reading room
25. reading area for lecture

University: Hu'nan University
Designer: Liao Ruowei
Tutor: Yan Xiangqi
Course Name: CZech Technical-Hu Nan University Joint Design Studio
Finished Time: Jan., 2015
Exchange Institute: Czech Techincal University School of Architecture

Second Prize

49

[CITY OF BLINDING LIGHTS]

Community Library in Historical District of Changsha

The idea is developed from the fabric of the old city. The library is located in a historical block (Chao-Zong st.), thus I choose [City Memory & Fabric] as my starting point, try to find something in everyone's common memory about the old city.

The library is an [abstract miniature city], with veiled imagines of people's activities can be seen through translucent glass curtain wall. Visitors can not only read, but also wander through the old lanes and experience traditional activities.

Generalize [the old Changsha's fabric] into several sorts, use as reference of the location of the functional groups in the buildings. While city shows independence of buildings in block, in this design, each functional group shall have a independent block and a single function (such as reading room for one particular theme, or activity room for one item).

CITY MEMORY

fabric of the old block | buildings around the site | function grouping | districts and main roads

formation inside the group | transitional space and device | carry on the fabric | [city] integrate into the city

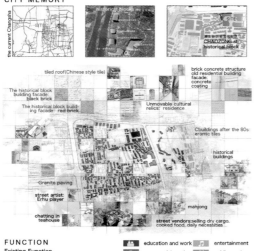

FUNCTION

Existing Function
- education and work
- entertainment
- commercial service
- public service

Function Addition according to previous research: Traditional Activity Group

"Traditional activities are more than mahjong". Hoping to arouse the memory of the old days, enrich folks' entertainment and communication.

Function Arrangement
- Reading&Lecture
- Cafe
- Activities
- Communication

Function Grouping

the service	logistics and staff		
the serviced (within the group: the way to get information)	[Independent search]: through the media — reading room, study room	[passive acceptance]: through the speaker — training classrooms, lecture hall	[interactive heuristic]: through communication — traditional activity group

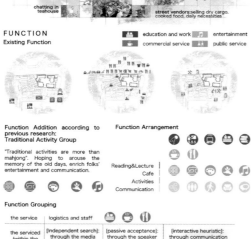

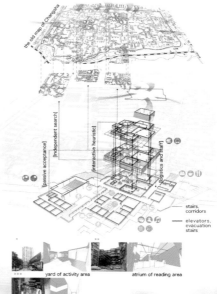

- yard of activity area
- atrium of reading area
- corridor of lecture area

stairs, corridors
elevators, evacuation stairs

[passive acceptance] [independent search] [interactive heuristic] [logistics and staff]

Space formation
[response to context: folks nearby and the old Changsha memory]
Should not be figurative but abstract, giving affinity, attribution, the sense of cultural identification. It should be public perceived.
=> how : abstract the old Changsha City map for space formation.

the distinction of commercial, residential, cultural and education area.
Groups are separated from each other, and a wide corridor in between.

a plurality of gate => add more site entrance, improve accessibility

The main road and secondary roads. => internal connection between the groups, it is also the open activity space.

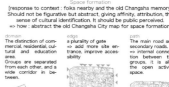

the marker — famous or memorable buildings... => referred for deformation

node (the typical city node) => referred for the plan arrangement inside the groups

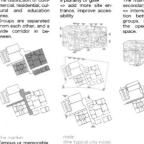

Central radiation: cohesion — [interactive heuristic]

Eccentric radiation: centrifugal scattering — [Independent search]

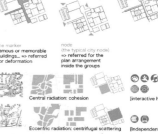

Straight street; efficient allies — [passive acceptance] [logistics and staff]

NORTH ELEVATION   WEST ELEVATION

作品名称：光炫之城——长沙历史街区中的社区图书馆
City of Blinding Lights: Community Library in Historical District of Changsha

二等奖

院校名：湖南大学
设计人：徐嘉韵
指导教师：蒋甦琦，邓广
课程名称：建筑设计A3
作业完成日期：2015年01月
对外交流对象：斯洛文尼亚卢布尔雅那大学建筑学院

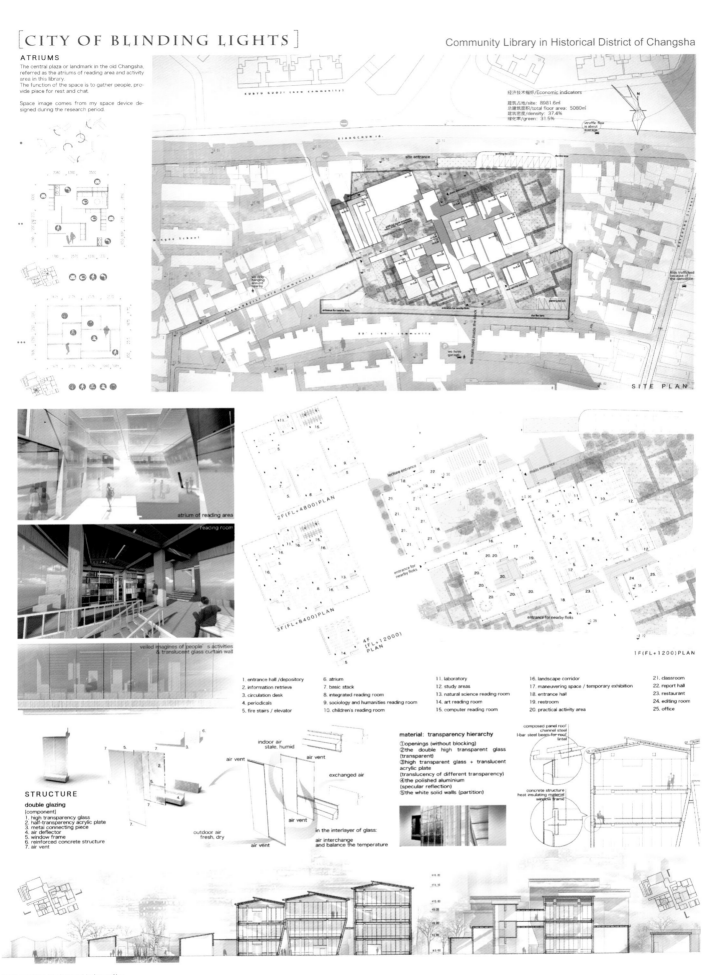

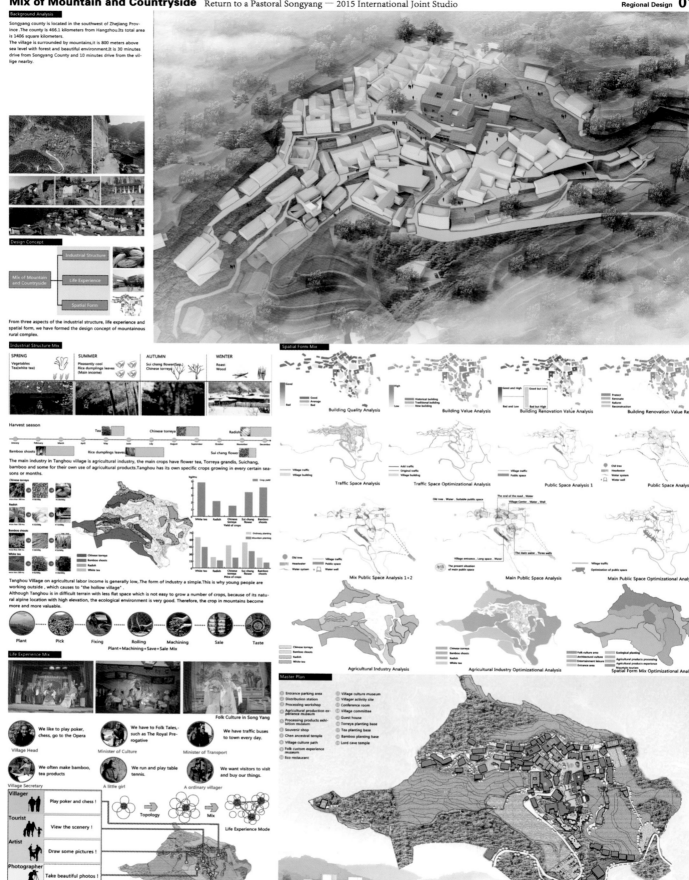

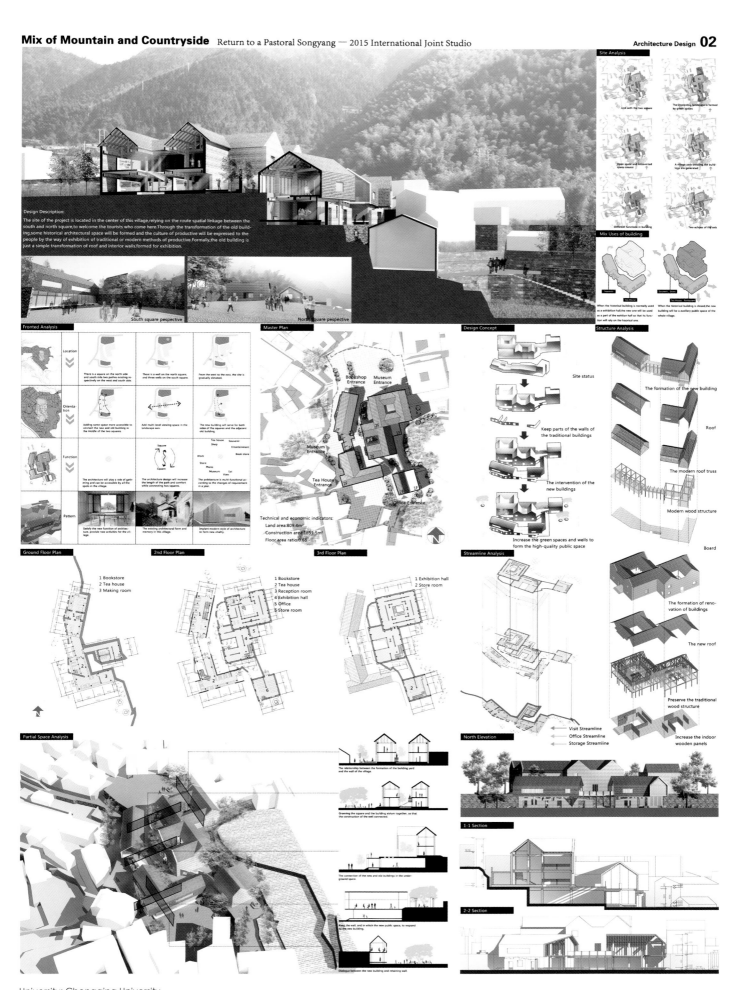

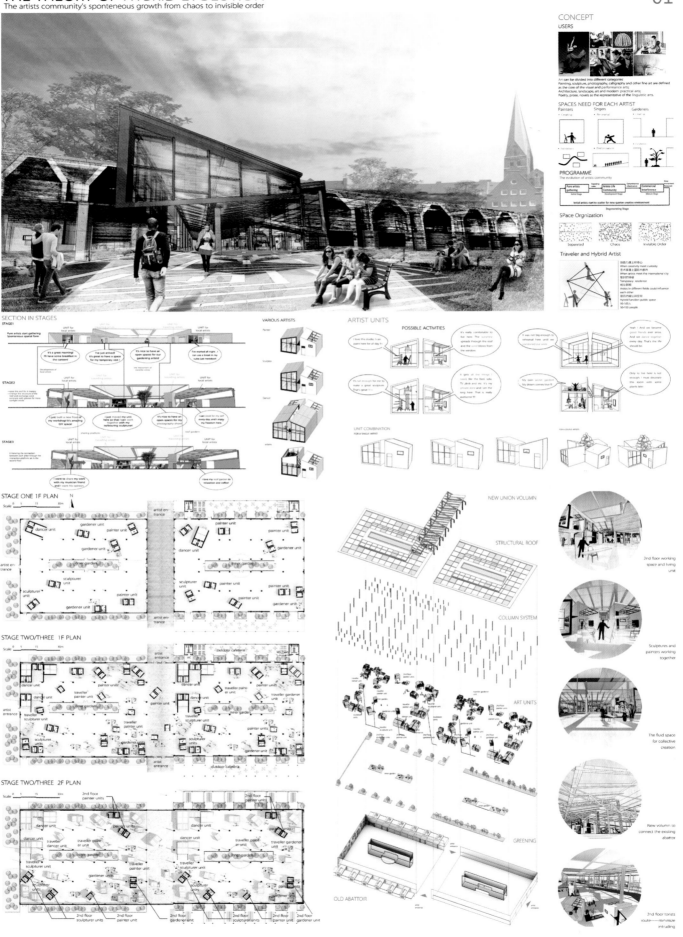

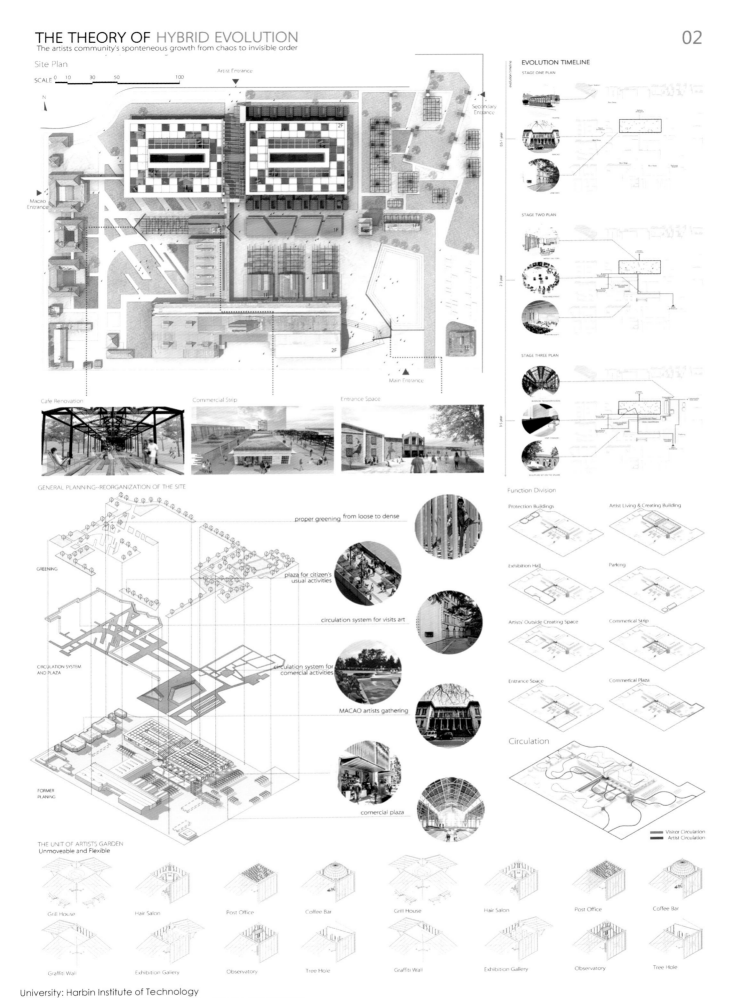

THE THEORY OF HYBRID EVOLUTION
The artists community's spontaneous growth from chaos to invisible order

University: Harbin Institute of Technology
Designer: Shi Yuqing, Zheng Yunchao, Guo Wenjia, Wu Xuefeng
Tutor: Meng Qi, Liang Jing, Nadia Bertolino
Course Name: International Joint Workshop
Finished Time: Jul., 2015
Exchange Institute: The University of Sheffield

Third Prize

MY TINY NEW YORK
A vertical market tower at Grand Central Terminal

01

Intepretion of NY City

Diverse culture

- own studio
- fruit business
- start APP store
- home-farm fruits
- healthy life style
- homemade business

Diverse peoples & demands

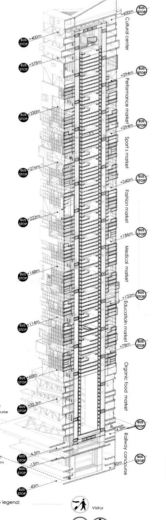

Intepretion of the Site

New York as a world-class financial centre, its high density is a response to the concentration of people from all around the world, with diverse backgrounds. Thus tiny and diverse space becomes a distinct feature of the city, which brings vibrant and convenient urban life. Our proposal is in turn with this characteristic and seeks new possibilities to **accommodate and facilitate diverse culture and people**.

The site is located beside the Grand Central Station, which is one of the biggest commuting stations worldwide. We noticed an opportunity to contribute to this situation in the form of **a vertical market**, which can also stimulate a new possibility for the typology of high rise buildings. We want it to be multi-cultural, mix-used, populist and as accessible as traditional streets. People from different cultural background can **find their necessities here**.

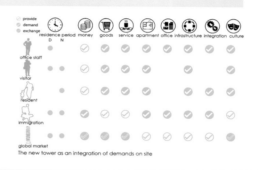

The new tower as an integration of demands on site

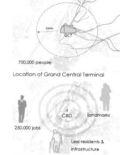

Location of Grand Central Terminal

700,000 people

250,000 jobs — CBD, landmarks

Less residents & infrastructure

Demands of people on site

System Evolution

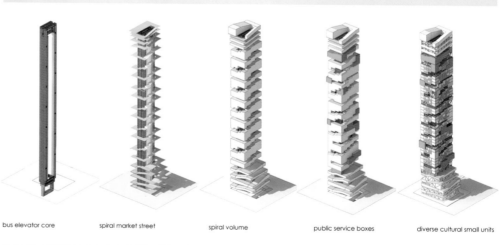

bus elevator core | spiral market street | spiral volume | public service boxes | diverse cultural small units

Growth of Small Units

at first | half year later | one year later | two years later | three years later

---

作品名称：我的微纽约——位于中央车站的垂直市场大楼
My Tiny New York : A Vertical Market Tower at Grand Central Terminal

三等奖

院校名称：同济大学
设计人：程思，李祎喆，张谱，赵音甸
指导教师：谢振宇，王桢栋，谭峥
课程名称：同济大学—CTBUH—KPF联合设计课程
作业完成日期：2015年06月
对外交流对象：世界高层建筑与都市人居学会（CTBUH），KPF建筑设计事务所

MY TINY NEW YORK
A vertical market tower at Grand Central Terminal

Our design integrates the underground space of the station with **proposed facilities assembled vertically** along the building, which are chained together by **a spiral pedestrian market** running from the bottom to the top. The system of facilities is predefined: a food court at the bottom shared by the station, a cultural center on the top to pull up pedestrian, and an organic food market, a set of public classrooms, a medical center, a fashion market, sport clubs, a theater center and indoor parks are scattered in between. Their accessibility and efficiency are ensured by a **"bus elevator" system**, which runs automatically on schedule and stops every 54m where the facilities reside. Tiny and flexible spaces around are left for lease and **spontaneous growth**, which is expected to be compatible with certain facility. Residential spaces are also provided, usually strongly associated with the entrepreneurs. Due to the **small scales**, the rent is more **affordable**, making the **cultural mixture** possible, and allowing for **self-regeneration**.

We envision the future situation that small and diverse entrepreneurs agglomerate around certain facilities, within them are different **cultural clusters**. For example, around the medical center, there would be entrepreneurs of all kinds of medical products, including those of Chinese herbal medicine, of Korea traditional medicine, Italian dental clinic and so on.

People brought by trains flow from Grand Central Station to this vertical market, guided by **the spiral pedestrian system** or carried by the **"bus elevator"**, and with those facilities as landmarks, find different **goods and service** as easily and effectively as on a traditional street. And different cultures manifest themselves in this building.

Top——An Open Cultural Center

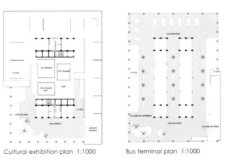

Cultural exhibition plan 1:1000   Bus terminal plan 1:1000

Middle——An Spiral Market Street

Typical public service plan 1:1000   Typical small units plan 1:1000

Bottom——A Global Food Market

Circulation: LIRR visitors   Circulation: North Metro&Subway visitors

Circulation: Cars   Circulation: Goods

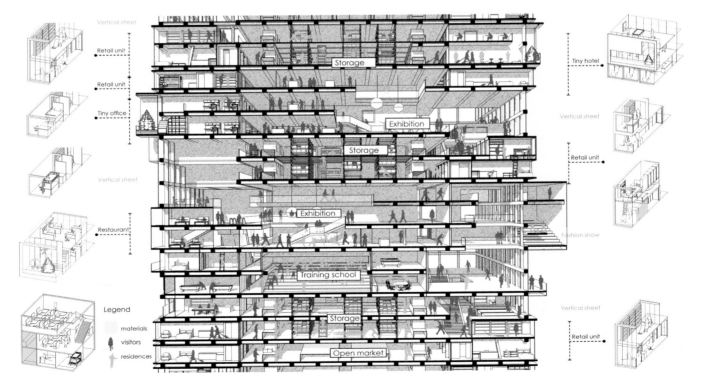

University: Tongji University
Designer: Cheng Si, Li Yizhe, Zhang Pu, Zhao Yindian
Tutor: Xie Zhenyu, Wang Dongzhen, Tan Zheng
Course Name: Tongji-CTBUH Jonit studio
Finished Time: Jun., 2015
Exchange Institute: Council on Tall Buildings and Urban Habitat (CTBUH), KPI

Third Prize

URBAN CARNIVAL | OFFICE BUILDING DESIGN OF CULTURAL CREATIVE INDUSTRY

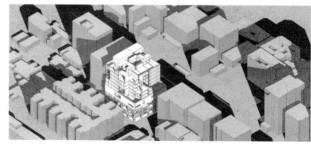

URBAN RESEARCH

SITE ANALYSIS
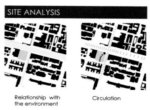

Relationship with the environment | Circulation

Open Space | Function

PROBLEMS AND IDEAS

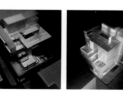

作品名称：城市嘉年华
Urban Carnival

三等奖

院校名：西北工业大学
设计人：方帅
指导教师：赖怡成，杨卫丽
课程名称：都市市集及办公大楼设计
作业完成日期：2015年06月
对外交流对象：淡江大学建筑学系

URBAN CARNIVAL | OFFICE BUILDING DESIGN OF CULTURAL CREATIVE INDUSTRY

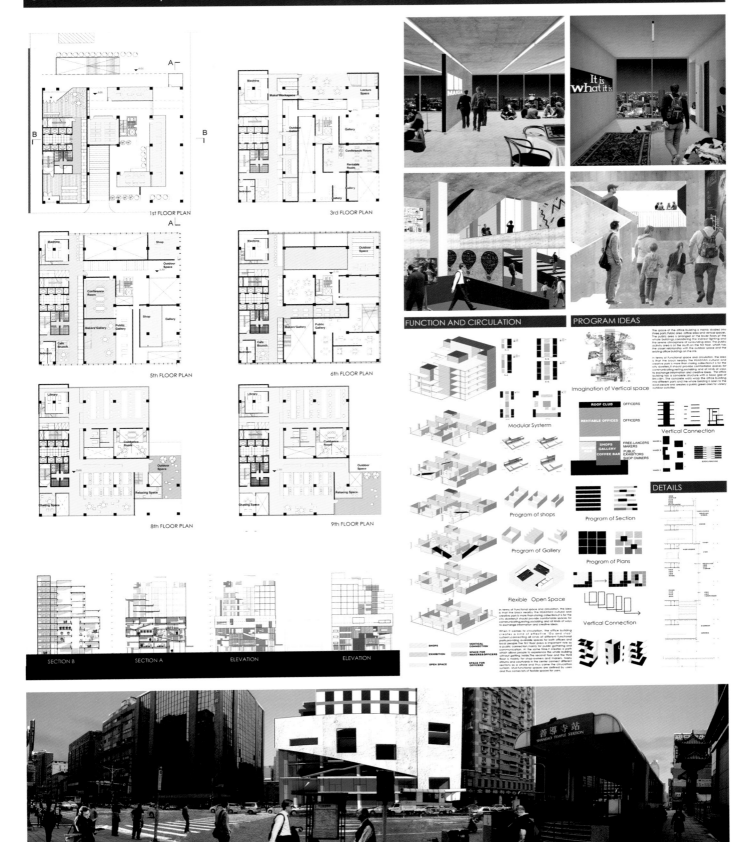

University: North western Polytechnical University
Designer: Fang Shuai
Tutor: Lai Yicheng, Yang Weili
Course Name: Urban Market and Office Building Design
Finished Time: Jun., 2015
Exchange Institute: School of Architecture, Tamkang University

Third Prize

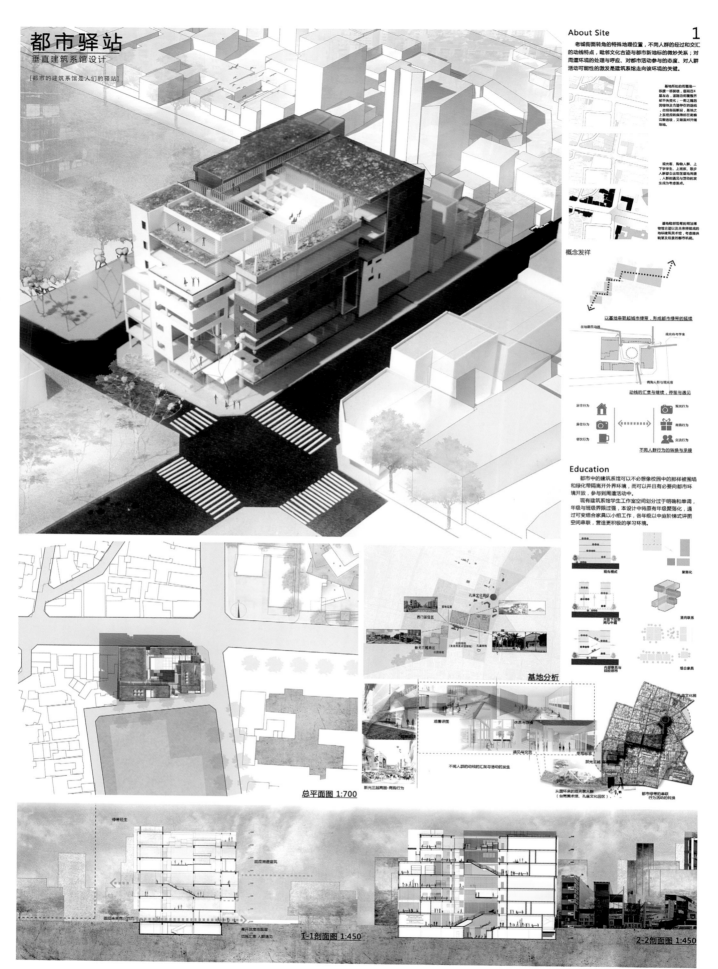

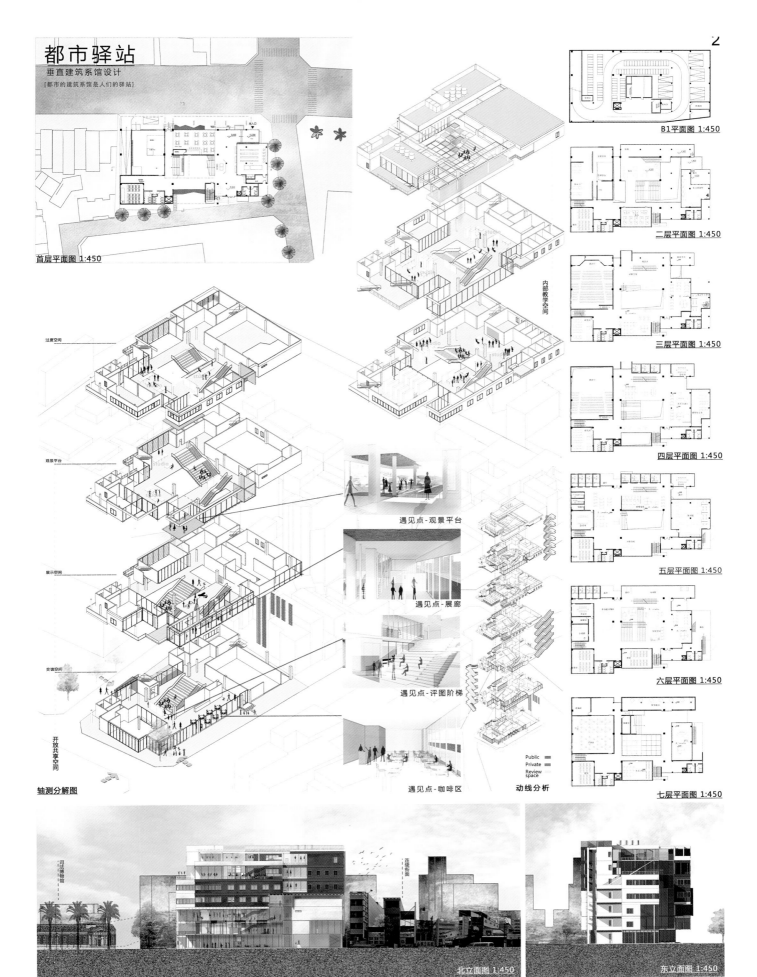

University: Dalian University of Technology
Designer: Wen Lianghan
Tutor: Yan Mao Chang, YuHui, Wang Shiyuan
Course Name: Vertical campus design
Finished Time: Apr., 2015
Exchange Institute: Cheng Kung University Department of Architecture

Third Prize

REDISCOVER THE VALUE HIGH UP IN NEW YORK
3-Dimentional City Space Stimulator

01

Observation
The After the site research, we discovered that though, there is a 3-dimensional space in NY city, there is no 3 dimensional life. The vitality of city remains near the ground, and the space high up become the ends of the city.

Proposal
we propose a new transportation infrastructure network high up in the sky (in the case of NY, the cable system) with the highrise, adding a force on top of the city. When the points high up in the sky are connected to form a network of people's daily life, the value of them are raised, and new possibilities for life are brought to the upper level of the city. This radical system supplement the grid system.

PRESENT NEW YORK SPACE VALUE

CONCEPT
What if the high-rise makes use of its advantage of height, providing an opposite force high up in NY, thus increasing the values at both ends?

WHERE TO BUILD?
We hope to establish our tower of vertical transportation infrastructure to form a three-dimensional network. Based on existing analysis, we select the points with the most close relationships with our site (the grand central), including the place where commuters, visitors, business travelers, and consumers come. Afterwards we overlap all these points and discover the points where the connections are truly needed.

Although New York City has a three-dimensional space, but it dose not have a 3D city life. All vigourous urban life are mainly concentrated in the ground layer and nearby underground subway, namely lower city level network.

When we arrived at the site, we found that the grand central terminal is the present situation of New York. Near the central station ground layers of transportation infrastructure is the also the starting surface of city life.

New York is a city of commuters. A lot of people going to work in New York can't take the bus and nearly half of residents go to work by car from brooklyn to queens, as the public transportation you can choose is limited. If one selects subway, he has to transfer several times.

While supplying a transportation network above the city, We are also creating some skypath for and more than commuters.

VERTICALLIY UNEVEN VITALITY

COMMUTER CITY

BEFORE / **AFTER**

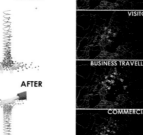

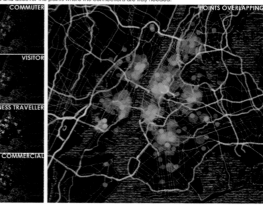

COMMUTER / VISITOR / BUSINESS TRAVELLER / COMMERCIAL / POINTS OVERLAPPING

INCREASED VALUE
NETWORK
After establishing the urban hub level, we set up the second hub level cable network in the surrounding area. This system is divided into two levels, the connection between the hub towe and the connection from hub to the surrounding high-rise tower, which corresponds to different height and different range of radiation. Eventually we build a three-dimensional transportation infrastructure network.

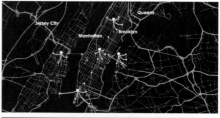

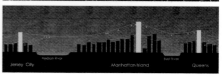

作品名称：垂直价值激发器
Rediscover the Value High up in New York

院校名称：同济大学
设计人：郑攀，牟娜莎，承晓宇，邹梦昊
指导教师：谢振宇，王桢栋，谭峥
课程名称：同济大学—CTBUH—KPF联合设计课程
作业完成日期：2015年01月
对外交流对象：世界高层建筑与都市人居学会（CTBUH）、KPF建筑设计事务所

三等奖

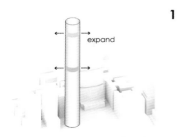

1

Add the transportation hubs (one city hub and one regional hub) to tall building and we get two important nodes. They expand with the urban one a little bigger than the regional one.

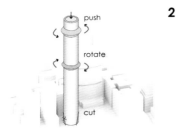

2

The hubs rotate to suit the cable car lines from different directions. Then we push the top to get the upper terrace and cut the bottom to leave public ground space.

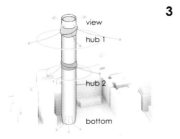

3

Here we get the two hubs and form the traffic network. The top area has a good view of the city and the bottom area is well connected with Grand central station.

4

The facade also varifies along the two hubs and we use vertical lines to emphasis the disturbance at the hubs.

1 Value & Vitality (nodes)

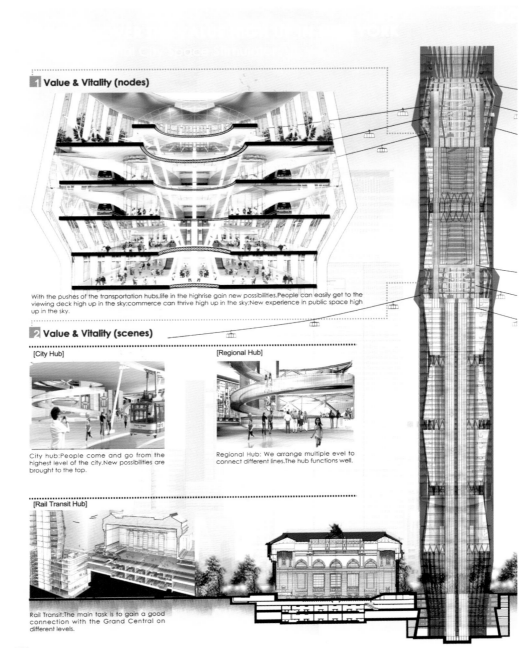

With the pushes of the transportation hubs, life in the highrise gain new possibilities. People can easily get to the viewing deck high up in the sky; commerce can thrive high up in the sky; New experience in public space high up in the sky.

2 Value & Vitality (scenes)

[City Hub]
City hub: People come and go from the highest level of the city. New possibilities are brought to the top.

[Regional Hub]
Regional Hub: We arrange multiple evel to connect different lines. The hub functions well.

[Rail Transit Hub]
Rail Transit: The main task is to gain a good connection with the Grand Central on different levels.

Roof Platform | Hotel | Hotel | Transit Hub | Transit Hub | Office | Office | Transit Hub | Commercial | ground public space

University: Tongji University
Designer: Zheng Pan, Mu Nasha, Chen Xiaoyu, Wu Menghao
Tutor: Xie Zhenyu, Wang Zhendong, Tan Zheng
Course Name : Tongji-CTBUH Joint Studio
Finished Time: Jan., 2015
Exchange Institute: Council on Tall Buildings and Urban Habitat (CTBUH), KPF

Third Prize

WEAVE REPAIR UPDATE
Return to a Pastoral Songyang 2015 International Joint Studio

Regional Design 01

· Current Situation
Songyang an ancient county of more than 1800 years history. It is located in Lishui City Zhejiang province. There are multiple traditional natural villages in Songyang, which are historic and distinctive.

The buildings in the old street are badly damaged. Many of them are disuse or inactivity, without repairing.

Heavy traffic caused by the narrow streets, short of parking lots, make it hard to get into the old street.

· CONCEPT
Starting from the current situation, generating related strategies by the combination of internal and external means, thereby rebuilding the old street from material and spiritual aspects.

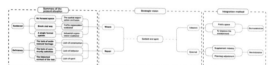

· DIAGRAMS

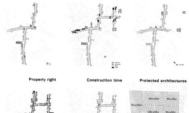

Property right | Construction time | Protected architectures

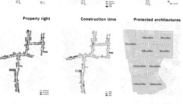

Architecture scene | Construction layer | Streets scale

· FORMS of COURTYARD

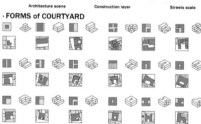

· CONSTRUETION PRINCIPLE
Single building construction principle
To avoid the construction of large-scale cluttered, box type construction, the construction scale to fit for human scale, building height shall be refer to the width of the streets, the confining sense, good floor size and the building volume control. Ensure that the streets of the building coherent, clear the street interface.

Group building construction principle
The harmony of group building pays attention to the harmony of the same area group building, and the mutation of the architectural style, consideration and modeling should not be caused.. Maintain the pattern of block courtyard, continue block texture.

· STRATEGYS

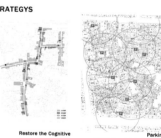

Restore the Cognitive | Parking | Restructuring Forms

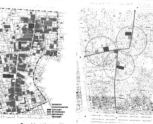

Combing the traffic | Business | Traffic System

Layers

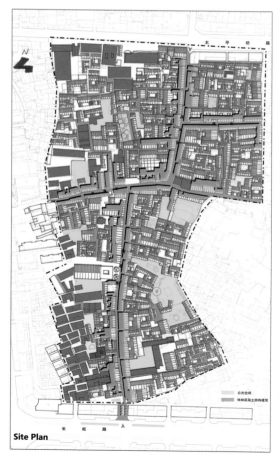

Site Plan

作品名称："织补"——基于历史文脉重塑的松阳老街更新设计
Weave Repair update

三等奖

院校名：重庆大学
设计人：李忠明，马汀，王嘉睿
指导教师：邓蜀阳，阎波
课程名称：回归田园松阳：2015国际五校研究生联合设计
作业完成日期：2015年05月
对外交流对象：日本早稻田大学，香港大学

CULTURE CENTER
Return to a Pastoral Songyang 2015 International Joint Studio

Architecture Design 02

PROJECT POSITION
With the continuous development of social economy, people living standard.To improve the traditional architectural space, has beenunable to meet the people's daily.Life, therefore, we must make appropriate adjustments to theexisting space. Change, to adapt to the development of the times. The city environment in order to meet the needs of the residents of the city.The necessary adjustments and changes, thatis the essential stage of city renewal,Is the accelerated process of industrialization, the economic structure change based on. Profound changes in social all-round development of the macro background. Function replacement of material space and human-space history block.

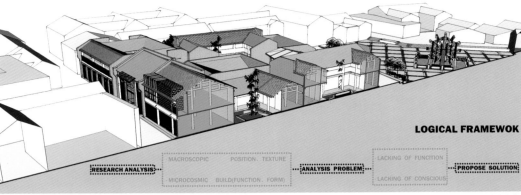

LOGICAL FRAMEWOK

RESEARCH ANALYSIS — MACROSCOPIC POSITION.TEXTURE / MICROCOSMIC BUILD(FUNCTION.FORM) — ANALYSIS PROBLEM — LACKING OF FUNCTION / LACKING OF CONSCIOUS — PROPOSE SOLUTION

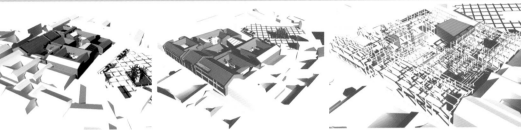

- **TECHNICAL DRAWINGS**

First floor plan Second floor plan

A-A Profile

West elevation

- **COGNITIVEFIELD**
- **CONCEPT GENERALIZATION**

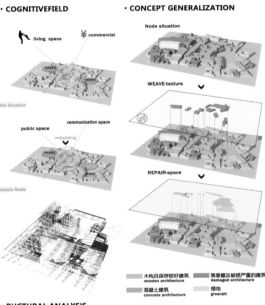

- **RUCTURAL ANALYSIS**

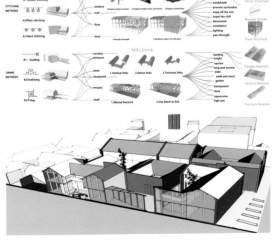

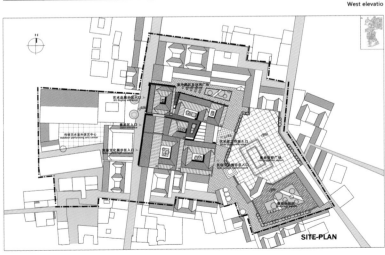

SITE-PLAN

University: Chongqing University
Designer: Li Zhongming, Ma Ting, Wang Jiarui
Tutor: Deng Shuyang, Yan Bo
Course Name : Return to a Pastoral Songyang-2015 International Joint Studio
Finished Time: May., 2015
Exchange Institute: Waseda University, The University of Hong Kong

Third Prize

ARCHITECTURAL DESIGN IN DEVELOPING COUNTRIES 1
PROTOTYPE ARCHITECTURE FOR EMERGENCIES

This design is based on the remould of an abandoned gymnasium in old city, through surveying and researching. I designed this transformed architecture for exhibitions, meetings, and food trading center. The food for trading is considering the speciality of the city or trade needs around the city.

The function of each space is not defined. Instead, it is the people who can choose and enjoy space freely by the timeless of architecture.

The form design of architecture stives to integrate into the church around it by mainly considering original spatial structure, massing alignment with surrounding buildings, and traffic connection.

PICTURE 1

PICTURE 2

PICTURE 3

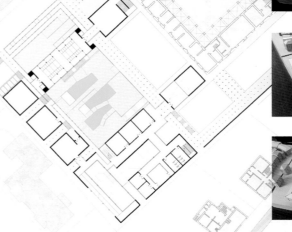

GROUND FLOOR

SECOND FLOOR

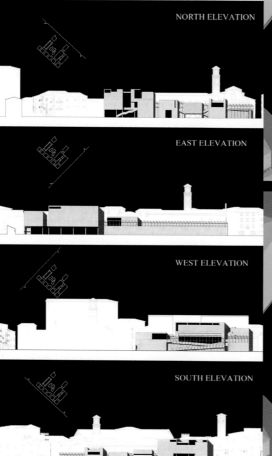

NORTH ELEVATION

EAST ELEVATION

WEST ELEVATION

SOUTH ELEVATION

ISOMETRIC A

ISOMETRIC B

作品名称：旧城改造 fidenzal

院校名：大连理工大学
设计人：李东祖
指导教师：Domenico，李冰，范悦
课程名称：architectural design studio 1
作业完成日期：2015年07月
对外交流对象：意大利米兰理工学院

三等奖

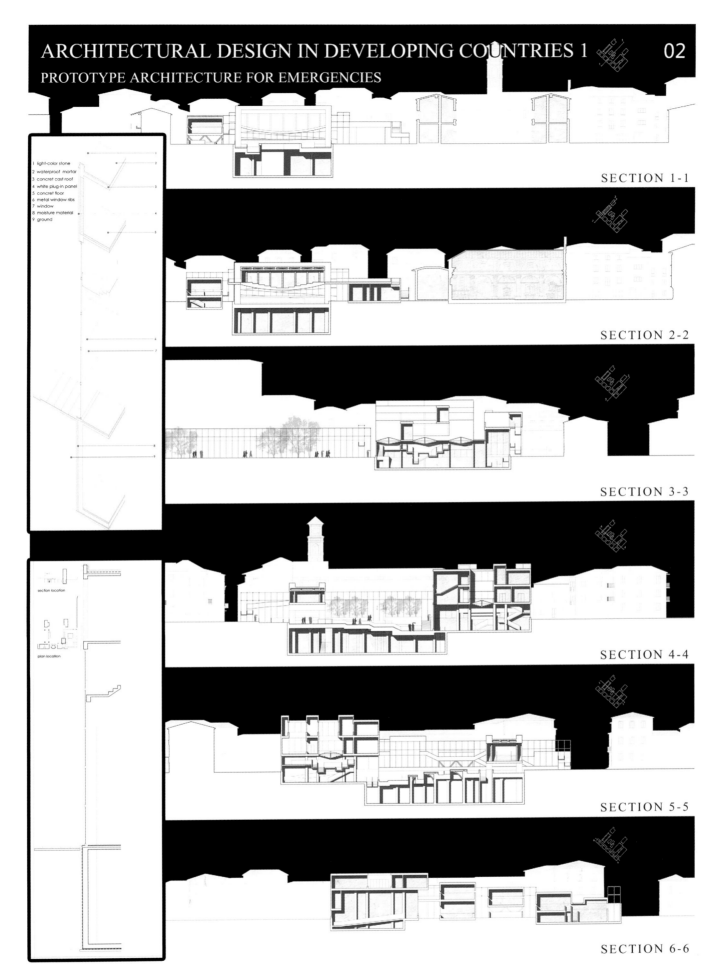

ARCHITECTURAL DESIGN IN DEVELOPING COUNTRIES 1
PROTOTYPE ARCHITECTURE FOR EMERGENCIES

SECTION 1-1

SECTION 2-2

SECTION 3-3

SECTION 4-4

SECTION 5-5

SECTION 6-6

University: Dalian University of Technology
Designer: Li Dongzu
Tutor: Domenico, Li Bin, Fan Yue
Course Name : Architectural Design Studio
Finished Time: Jul., 2015
Exchange Institute: Politecnico Di Milano

Third Prize

SEQUENCE - INHERITIED

序列·传承——唐轴线小雁塔地段城市设计

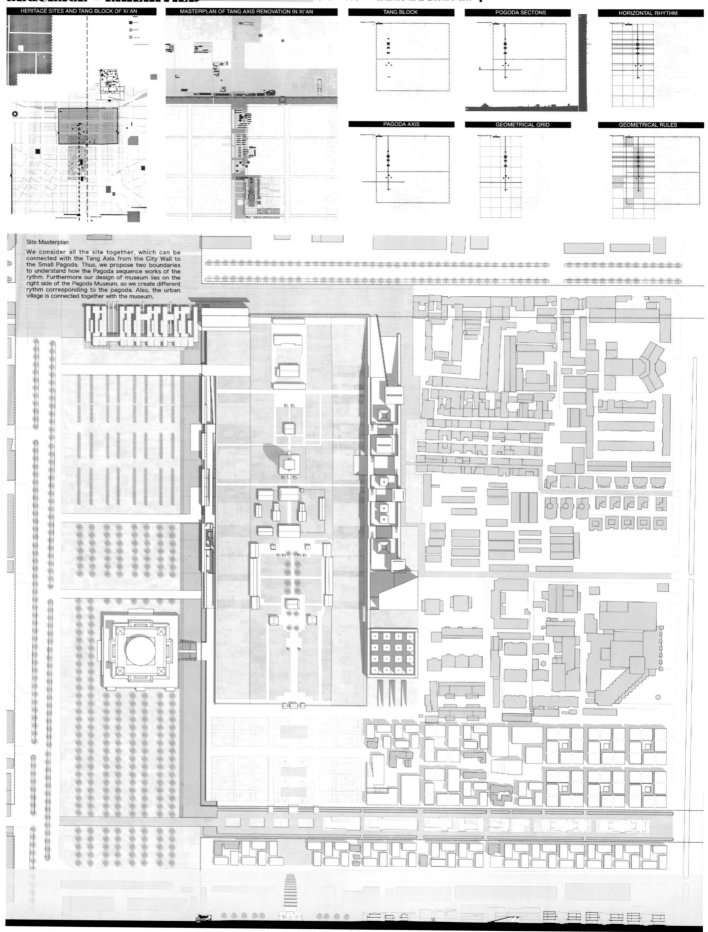

作品名称：序列·传承——唐轴线小雁塔地段城市设计
Sequence-Inheritied

院校名称：西安建筑科技大学
设计人：卢肇松，王龙飞，李露昕，Claudia Grossi, Giulia Guida, Fillipo De Rosa
指导教师：李昊，常海青，鲁旭，李焜，laura Anna Pezzetti, Carlo Palazzolov
课程名称：2015西安建筑科技大学—意大利米兰理工大学国际联合工作营课程教学
作业完成日期：2015年05月
对外交流对象：意大利米兰理工大学建筑学院

三等奖

University: Xi'an University of Architecture and Technology
Designer: Lu Zhaosong, Wang Longfei, Li Luxin, Claudia Grossi, Gulia, Fillipo De Rosa
Tutor: Li Hao, Chang Haiqing, Lu Xu, Li Kun, Laura Anna Pezzetti, Carlo Palozzolo
Course Name: 2015 Polimi-XAUAT International Workshop
Finished Time: May, 2015
Exchange Institute: College of Architecture, Politecnico di Milana

Third Prize

■ 项目区位 LOCATION

一方天地 西安渭桥遗址区博物馆设计
ONE SQUARE WORLD_XI'AN WEI BRIDGE HERITAGE SITE MUSEUM

■ 渭河桥历史功能 FUCTION IN HISTORY

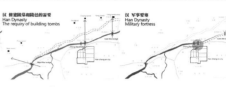

横门桥 连接宫殿	中渭桥 连接安陵与长陵	秦 连接渭河两岸宫殿	汉 丝绸之路上的第一座桥梁	汉 杜陵霸陵和阳陵的需要	汉 军事要塞
Heng Bridge, connect the palaces	Middle Wei Bridge connect An tomb and Chang tomb	Qin Dynasty connect north and south	Han Dynasty As the first bridge on silk road	Han Dynasty The requiry of building tombs	Han Dynasty Military fortress

■ 现状环境分析 CURRENT ENVIRONMENTAL ANALYSIS   ■ 设计说明 DESIGN DESCRIPTION

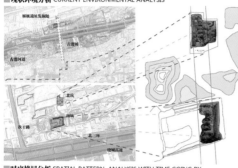

西安渭桥遗址博物馆位于西安市未央区西席村北的厨城门一号桥遗址上，为保护与展示古渭河桥遗址而建立。

2014年因村民挖沙暴露出木桩，遗桥被确定为古桥遗存并进行抢救性挖掘。桥梁东西宽约15.4米，南北长880米，挖掘出包括桥板、桥舟、古船等重要遗迹。据考古分析，渭桥不仅是同时期全世界最大的木梁柱桥，也是陆上丝绸之路的第一座桥梁；同时也是历史上一系列重大历史事件的原生地；更由于渭河多次改道，渭桥的发现对于历史时期渭河变迁以及关中地区自然环境的变化具有重要研究价值。

渭河是关中地区的母亲河，孕育了灿烂的中华文明，秦汉时期更是以渭河作为城市的重要依托。渭河作为通行的重要通道，历史地位不容忽视。沧海桑田，渭桥地亦重现于世，其建于贫沙若泥之中业于数千年的的古遗迹的佛身处另外一个时空。博物馆的设计概念即是锁在这一个永久远的时间点，造取河岸交界处遗留的浸润的时分，在保护遗址同时，让人们体验暮至千年，罗马奥汉时的壮阔画卷。而高处的桥栏模糊标识出即将随着考古工作的发展会逐步勾勒出渭桥的壮丽轮廓。

Xi'an Wei bridge hertige site museum is located in No. 1 bridge site, Chucheng Gate, Xixi village, District Weiyang, Xi'an city, which was built to protect and display the ancient bridge's heritage. In 2014, for the villagers sand-excavating exposed several stakes, and then was identified as the remains of ancient bridge and retrieved exploring. His Bridge was about 15.4 meters wide from west to west, north-south length of 880 meters, the excavation included bridge pier, consisting of shore, ancient boats and other sites. According to archaeological analysis, the Wei Bridge was the world's largest wooden beam-column bridge at that time, and the first bridge on the silk road; At the same time Wei Bridge is also a place where a series of major historical events happened; Moreover, because of the Wei river diversion for many times, the find of Wei bridge has had meaningful research value on Wei river's changing in history and the changes of natural environment in Guanzhong region. Wei river is the mother river of Guanzhong plainitia, gave birth to the Chinese civilization, the Qin and Han dynasties depended on Wei river to build splendid cities. Wei river, as an important water path, the historical position cannot be ignored. Nowadays, the heritage of Wei bridge reappeared, its buried in silt for thousands of years are like in a different time and space. Museum's mainly design concept that is fixed in that point, where the river flooded and finally covered Wei bridge. Protecting sites at the same time, let people experience through one thousand, feeling the spectacular images of Qin and Han dynasties. The long bridge simulation logo is with the development of archaeological work to draw the outline of Wei bridge gradually.

■ 时序格局分析 SPATIAL PATTERN ANALYSIS WITH TIME GOING BY

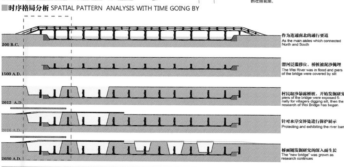  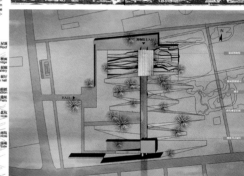

■ 渭透视效果图 SECTIONAL PERSPECTIVE

作品名称：一方天地——西安渭桥遗址区博物馆设计
One Square World : Xi'an Wei Bridge Heritage Site Museum

三等奖

院校名：西安建筑科技大学
设计人：高元丰，刘佳，罗婧，Mitch clarke，Jesse Jones，Jessica Philips
指导教师：常海青，苏静，同庆楠，吴涵儒，鲁旭，Albertus Wang，Martin Gold
课程名称：2015中美联合设计课程教学
作业完成日期：2015年06月
对外交流对象：美国佛罗里达大学建筑学院

Current Situation and Site Analysis

There is the intersection face to our site. Our site fronts on two streets. People mainly come from MRT station using Jing-Kan Road. Many cars come from high way using San-Min Road.

Concept

The "LOCK" of traditional house

For people frequently move one place to another, The housing units nowadays gather as much functions possible together into one single room.

These houses always comes along with so expensive price and low using frequency can be deemed as a kind of "lock".

The main concept "unlock" means a kind of reversal of traditional urban live pattern consists of many functional small units into one whole.

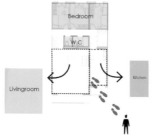

The "UNLOCK" of NEW house

Two Dimension of Unlock

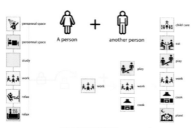

The "UNLOCK" of personnel compose

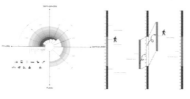

The "UNLOCK" of traditional timetable

Main People

Needs of Residents

Space Strategy

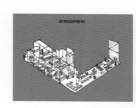

The psychological permanence of "home", either for single person or multiple-person households, is challenged cruelly by the changing nature of the contemporary society.

The moving social landscape is observed to be more intense and desired in cities. Mainly, it challenges traditional thoughts of home in four aspects: Mobility, Adaptability and Flexibility, Moving "socio-scape", Financial tightness

Space Strategy

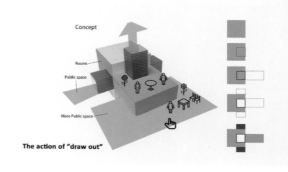

The action of "draw out"

Transition-
Intermediary Home 1/2

2015 Housing Workshop investigates the transitional aspect of contemporary homes. While we are holding our homes together, our homes are frequently at "intermediary" states. We are facing constant changes derived from life stage developments and their associated living contexts.

We move from one place to another because of changes of work conditions, family members' Jobs, partners, marriage status and birth of children, schooling, etc. People are forced to move for economic up-and-downs, career fluctuations, marriage status, and many unforeseen reasons.

We are looking for housing solutions to respond to these situations. We do like home stability yet knowing that we will move willingly or unwillingly to another place for another state of life. We quest for satisfying homes and adequate housing between moves.

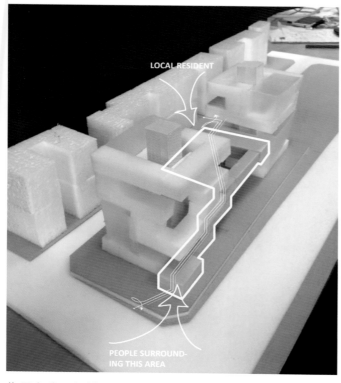

作品名称：解锁——过渡性住宅
Transition-Intermediary Home

三等奖

院校名：武汉大学
设计人：覃琛，米仓春菜，大荣桃世，阙暐桓，杨子兰，江能煜，郭锝湘
指导教师：张睿，胡晓青，黎启国
课程名称：台北东京都会住宅工作营
作业完成日期：2015年09月
对外交流对象：日本女子大学建筑系，淡江大学建筑系，逢甲大学建筑系

Site plan Configuration

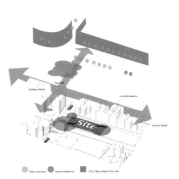

The urban site locates on the North east corner of San-Ming Rd. and Gen-Kan Rd. Taipei City, thus it needs to respond the road corner.

On the ground floor plan, we use paths and central square to ensure people nearby can easily travel through the lowel levels of the building, another leisure space will also be put into the ground floor plan.

Space Configuration

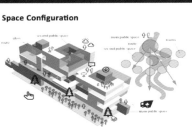

Our Scheme consists of three diffreent layers: One is main public space including the green space and lower levels parts.
one is secondary public space which undertakes the community activities. The third is private space which people live and stay.

View Relationship

The volumes overlap with each other create multiple ambiguity space relationship

The view relationship of several terrace extend people's view thus encourage the communication

Third Floor Plan

Fourth Floor Plan

Fifth Floor Plan

TYPE A for one　　TYPE B for three　　TYPE C for two

Transition-
Intermediary Home 2/2

Spatial Operation

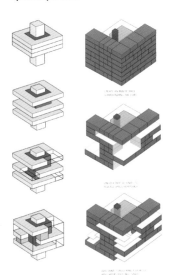

We break the "house" by drawing the functions which are not so indispensable like kitchens, living rooms and so on out of the housing units and convert them into social parts.
By drawing different space out of a solid volume, ambiguity space come into being, the solid part which is also the private part in the building become more compact with the adding of these public space

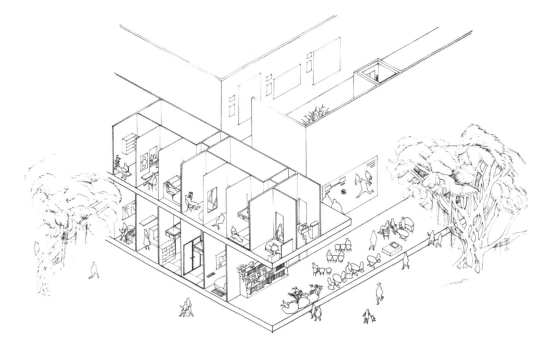

University: Wuhan University
Designer: Qin Chen, Yonekura Haruna, Oe Momoyo, Que Weihuan, Yang Zilan, Jiang Nengyu, Guo Dexiang
Tutor: Zhang Rui, Hu Xiaoqing, Li Qiguo
Course Name : 2015 JWU/TKU Taibei Tokyo Housing Workshop
Finished Time: Sep., 2015
Exchange Institute: Japan Woman's University, Tamkang University, Feng Chia University

Third Prize

STACK GARDEN

INTER-GENERATIONAL COMMUNITY: SERIOR LIVING, HEALTH AND WELLNESS, REHAB, CLINIC

Background:

Aging has became a global issue. Both the United States and China have been rapidly increasing elderly population. By 2020, both countries will have more than 20% of elderly population. Because of this situation How to design the physical environment for elderly to involved in the society and engage in physical social activities has became a challenge to both countries. In the U.S., different approaches have been explored, including independent living, assisted living, nursing home, and CCRC (Continuing Care Retirement Community). However, the above design, solutions are based on age-restricted communities. Older people are isolated from the other age groups.

In this case, we put our eyes to the elderly, we do the research about their living experience, daily activities, common behavior as well as health and wellness. we try to design a kind of inter-generational community that can satisfied with their daily life. convenient their activities, and try to give the elderly more special care and love.

Because of the elevation difference I come up with the ideal of stuck building which can give visitors Sense of depth.Then the design concept takes the idea from traditional Chinese garden, where visitors can experience the elegant surrounding as artificial space with natural elements, precisely arranged along the visiting route. The approach of porous green areas which combines fragments of open space together and provides a relatively gradual, cozy and humanized space system, The general layout is centered by a civc square and four buildings to two different edges. This also makes sure that public events can be held on the hall-enclosed square and in the meantime,outdoor space is shared evenly by visitors to four functional blocks.

MASTER PLAN

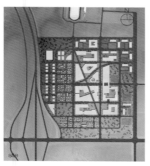
SITE

SITE FUNCTION ANALYSIS GRAPHICS

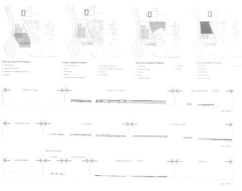

FUNCTIONAL RELATIONSHIP

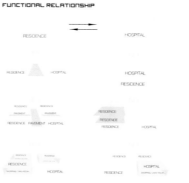

ALTITUDE ANALYSIS

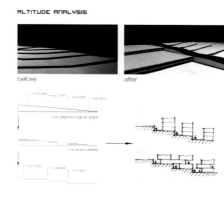
before / after

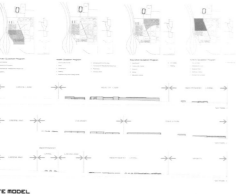

SITE MODEL

BIRD VIEW

EVOLUTION PROCESS

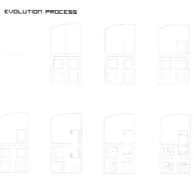

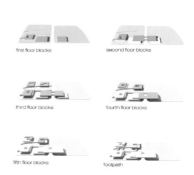
first floor blocks / second floor blocks / third floor blocks / fourth floor blocks / fifth floor blocks / footpath

---

作品名称：年龄混合型的老年居住、健康、医疗、康复社区规划与建筑设计
Inter-generational Community : Serior living, Health and Wellness, Rehab, Clinic

三等奖

院校名：南京工业大学
设计人：陈笑寒
指导教师：蔡志昶，方遥，蔡慧，Kent Spreckelmeyer
课程名称：联合毕业设计
作业完成日期：2015年06月
对外交流对象：美国堪萨斯大学建筑学院

STACK GARDEN

INTER-GENERATIONAL COMMUNITY: SENIOR LIVING, HEALTH AND WELLNESS, REHAB, CLINIC

PERSPECTIVE

PLAN

BLOCK EVOLUTION

FUNCTION ANALYSIS

TRAFFIC ANALYSIS

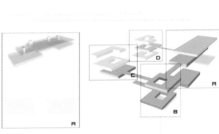

WALKING ROUTINE AND STAIRCASE

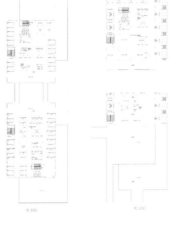

CONSTRUCTION DIAGRAM

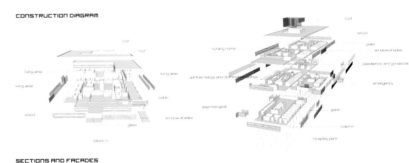

SECTIONS AND FACADES

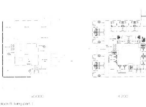

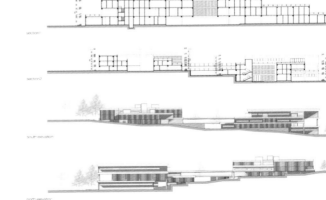

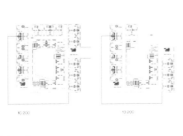

University: NanJin Tech University
Designer: Chen Xiaohan
Tutor: Cai Zhichang, Fang Yao, Cai hui, Kent Sprechelmeyer
Course Name : NanJing Tech University-Kansas University Joint Design Studio
Finished Time: Jun., 2015
Exchange Institute: Kansas University

Third Prize

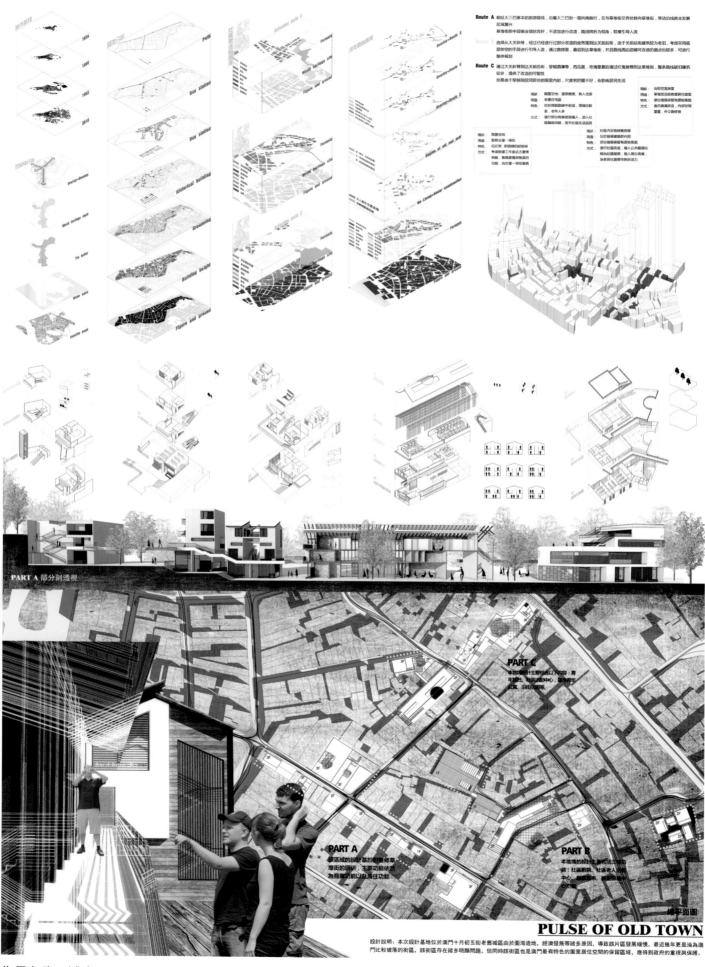

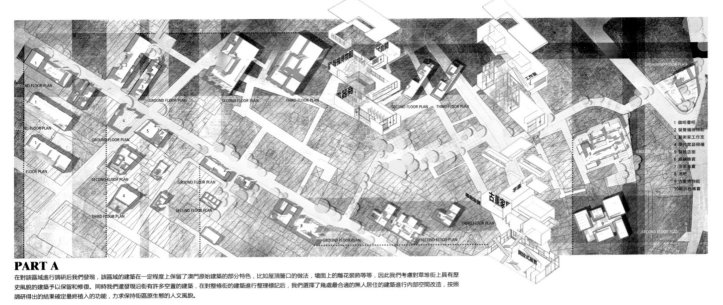

PART A

在對該區域進行調研后我們發現，該區域的建築在一定程度上保留了澳門原始建築的部分特色，比如屋頂簷口的做法，牆面上的雕花裝飾等等，因此我們考慮對草堆街上具有歷史風貌的建築予以保留和修復。同時我們還發現沿街有許多空置的建築，在對整條街的建築進行整理標記后，我們選擇了幾處最合適的無人居住的建築進行內部空間改造，按照調研得出的結果確定最終植入的功能，力求保持街區原生態的人文風貌。

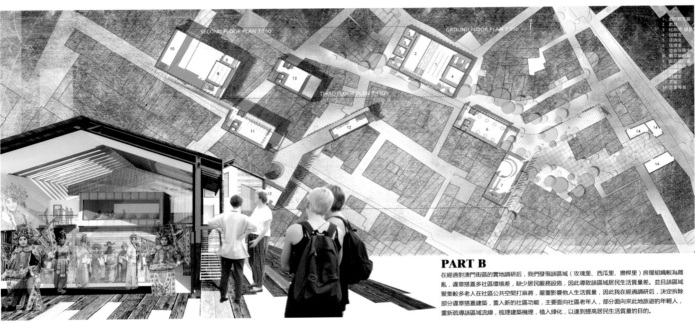

PART B

在經過對澳門街區的實地調研后，我們發現該區域（玫瑰里、西瓜里、擔桿里）房屋組織較為雜亂，違章搭蓋多社區環境差，缺少居民服務設施，因此導致該區域居民生活質量差。並且該區域聚集較多老人在社區公共空間打麻將，嚴重影響他人生活質量，因此我在經過調研后，決定拆除部分違章搭蓋建築，置入新的社區功能，主要面向社區老年人，部分面向來此地旅遊的年輕人，重新疏導該區域流線，梳理建築肌理，植入綠化，以達到提高居民生活質量的目的。

PART C

在經過調研后發現，該區域位於蜂里的出入口，有兩塊已經拆除建築的空地，可以被我們作建設新的適合社區功能的建築，並且該區域建築密度高，缺乏綠地，缺乏公共服務設施。另外該區域不容易被遊客發現，因此將其定義為不需極佳的地理位置也可以良好運營的青年旅社以及可租賃的單身公寓。

University: Huaqiao University
Designer: Shui Lu
Tutor: Cheng Li
Course Name: Redevelopment Master plan of Rua de Cinco de Outubro
Finished Time: Jun. 2015
Exchange Institute: the Land, Public Works and Transport Bureau of the Macao, Cultural Institute of Macao

Third Prize

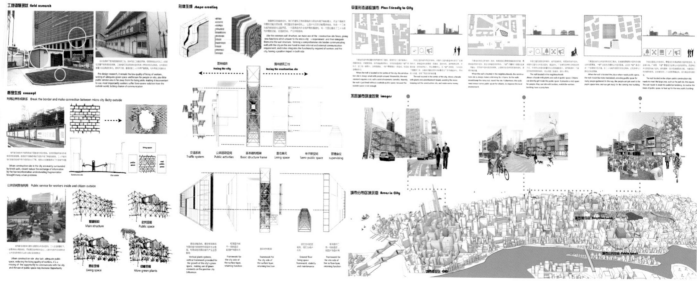

作品名称：城市建筑工地预制工人住宿
City Green "Wall"

三等奖

院校名：哈尔滨工业大学
设计人：李志斌，葛斐然，潘思傲
指导教师：连菲，唐家骏，Javier Caro Dominguez
课程名称：2015年哈尔滨工业大学联合设计课程教学
作业完成日期：2015年06月
对外交流对象：西班牙BAUM建筑设计事务所

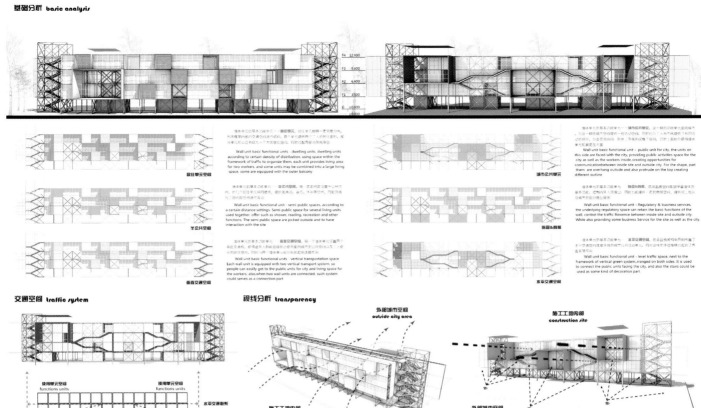
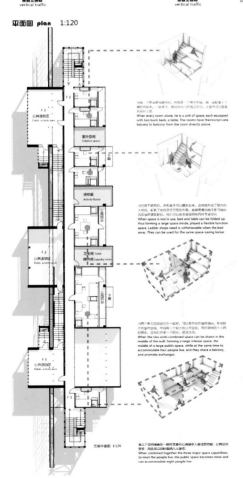
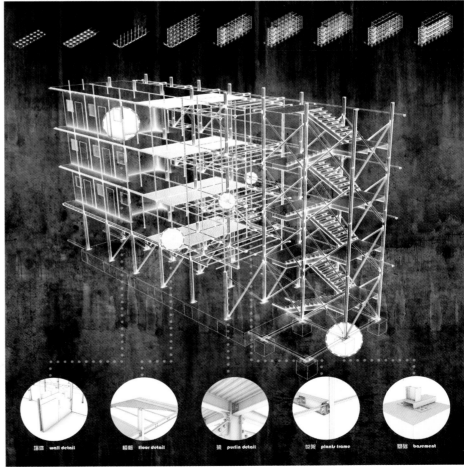

University: Harbin Institute of Technology
Designer: Li Zhibin, Ge Feiran, Pan Siao
Tutor: Lian Fei, Tang Jiajun, Javier Caro
Course Name: Harbin institute of technology-baum architecture design office
Finished Time: Jun., 2015
Exchange Institute: baum architecture design office

Third Prize

共 生

怡园历史街区更新城市及建筑设计

Design 01

该联合教学工作坊由东南大学、东京工业大学、同济大学和华南理工大学、苏州科技大学共同举行，以怡园历史街区更新城市及建筑设计为主题。本次联合教学的基地选址位于姑苏区内的古苏州城内，在人民路和干将路的交叉口，怡园以西。整个场地属于怡园历史文化街区内，场地周边建筑功能类型丰富，城市肌理明显，现拟将在场地内设计一个丰富的公共空间体系，既能保留苏州传统文化的丰富底蕴，又能提供给城市一个更新的机会，呼唤对空间价值的恰当理解和对城市发展的高度想象力。

本设计主题为"共生"，共生是指两种不同生物之间所形成的紧密互利关系。动物、植物、菌类以及三者中任意两者之间都存在"共生"。在共生关系中，一方为另一方提供有利于生存的帮助，同时也获得对方的帮助。在进行场地调研的时候，我们发现了该街区存在着丰富的建筑空间功能复合的情况，我们提出了共生的基本概念以此来恢复苏州历史街区的活泼的气氛。并通过对于基地的调研和对于城市的观察，找到对怡园历史街区的特定的城市条件下的一系列重要的共生的空间关系，并进行设计与规划。

城市中的共生现象 SYMBIOSIS IN CITY

Examples of Hybrid Space in a Single Building

Examples of Hybrid Space Among Multiple Buildings

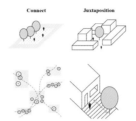

概念阐述 CONCEPTUAL ANALYSIS

共生是指两种不同生物之间所形成的紧密互利关系。动物、植物、菌类以及三者中任意两者之间都存在"共生"。在共生关系中，一方为另一方提供有利于生存的帮助，同时也获得对方的帮助。

树与空间 TREES AND SPACE

大树作为主要的公共活动场所，能够成为社区的核心。树木提供了阴凉，为人们创造了相对舒适的环境，人们常常自发地聚集在树下。
Trees could be the cores of the community, serving as a major public space. Providing shades from the sun, trees can create rather comfortable conditions and people intend to gather spontaneously under the trees.

Trees

There are a number of existing trees scattered around the site. Historically, objects such as these trees, wells, or bodies of water, often became the starting points and centers of communities. This project also proposes to use the existing trees as the core of the community, denoting public spaces around which the community can grow. In this way we can create new spaces while at the same time maintaining the memory of the district.

总平面图 THE SITE PLAN

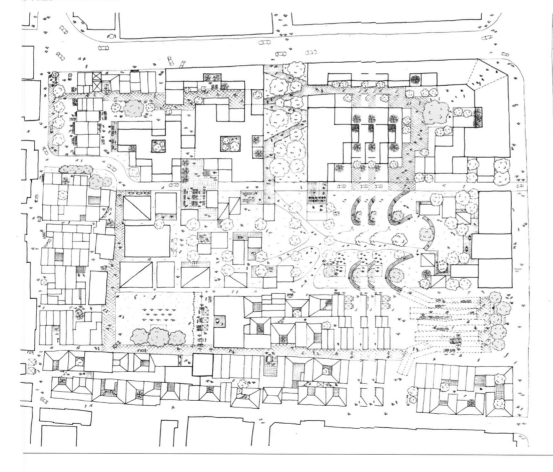

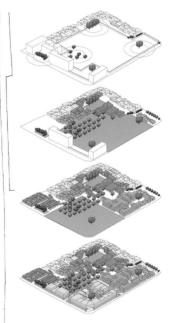

Time Lapse
The area is developed over time, centered around the growth points of the existing trees and based on the wishes of the residents. This gives flexibility to the residents to make their own space, though major functions such as the hotel and museum are predetermined with respect to the activities of the people.

- Residential
- Commercial
- Museum
- Community / Education
- Hotel

作品名称：共生——怡园历史街区更新城市及建筑设计
Symbiosis: Urban Design and Architecture Design of Joyous Garden Historical District

院校名：东南大学
设计人：刘怡宁，许健，吴俊熙，Joey LIPPE，Suzuki Aiko
指导教师：唐芃，葛明，奥山信一，王伯伟，王方戟，孙一民，胡莹
课程名称：中日五校联合工作坊
作业完成日期：2015年06月
对外交流对象：日本东京工业大学建筑学院

三等奖

SYMBIOSIS

Urban Design and Architecture Design of Joyous Garden Historical District

Design 02

The joint teaching workshops by the SEU,TIT, TJU and SCUT, USTS jointly held with grace historical block update urban and architectural design as the theme. Thebase of the joint teaching location is located in the ancient suzhou gusu area in the city, at the intersection of ganjiang road and renmin road, west of grace. Thisarea function type is rich, urban skin texture clear, now will design a rich in the field of public space system, which would hold suzhou rich traditional culture background, and can offer a chance to update the city, calling the appropriate understanding of space value and a high degree of imagination to the urban development. Symbiosis is a scientific term that typically refers to the long and intimate, mutually beneficial relationship betweentwo or more biological species. A phenomenon found in animals, plants, fungi, etc., each party in the relationshiprelies on the actions of the others for their mutual survival. We propose symbiosis as the fundamental concept forrestoring a healthy and lively atmosphere back to the historic district of Suzhou. By observing the behaviors ofhumans and buildings in the city, we can find a series of key symbiotic relationships in the urban condition.

局部效果图 PARTIAL RENDERING

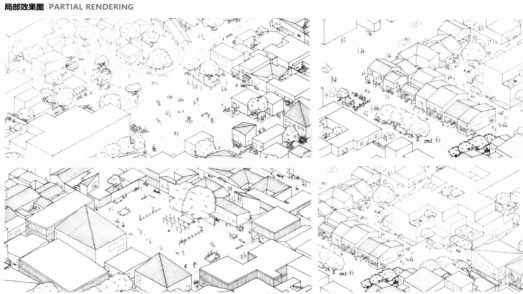

模型 THE MODEL

场地分体图 THE SITE ANALYSIS DIAGRAM

University: Southeast University
Designer: Liu Yining, Xu Jian, Wu Junxi, Joey LiPDE, Suzuki Aiko
Tutor: Tang Peng, Ge Ming, Shinichi Okuyama, Wang Bowei, Wang Fangji, Sun Yimin, HuYing
Course Name: Tokyo Tech-Tongji-Dongnan-SCUT-Suzhou Tech University Joint Design Studio
Finished Time: Jun., 2015
Exchange Institute: School of Architecture, Tokyo Institute of Technology

Third Prize

DIFFUSION

Diffusion on the City View

When the traditional housing area where the local live, and the tour spot where visitors flow, get close or overlapped.

Our site becomes special, for it has them overlapped.

We intend that diffusion is to offer local much convienience with public facility, and get visitors experience more into the real traditional street life.

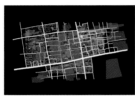
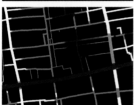

Pespective

Diffusion on Site
Scale Diffuse
Scale diffuse around our site. Buildings are composed of big and small scales.

Function Diffuse
Functions diffuse around our site. Commerce, museum and community center much related to the surroundings.

People Diffuse
Now the density of people decrease from metro to traditional area. Then we decide to have a stratage to diffuse in scale, function and people level.

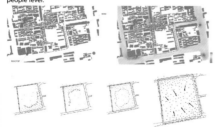

Design Strategy

Examples of Diffusion in Nod Square

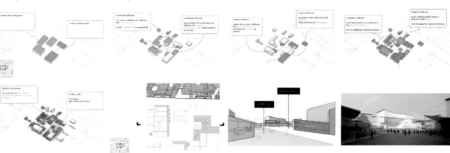

Facing the square, there're different functions.
So activities like cafe, music library, retail, studio, living happen around the edge of square. Outdoor cafe and "listen to KUN" square may have a relation.

Facing the open market, there're many functions combined with commerce.
There're silk display, handicraft display, museum here. They're all benifit for the goods related. Different special kinds of shops define an active outdoor market.

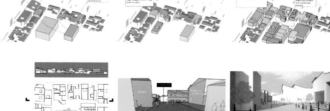

Facing the community open space, there're different functions.
There are shops and library around the plaza. Then we make a new relationship between traditional residencial area and community area.

Section

作品名称：扩散
Diffusion

三等奖

院校名：华南理工大学
设计人：陈碧琳
指导教师：孙一民，李敏稚
课程名称：五校联合工作坊
作业完成日期：2015年02月
对外交流对象：日本东京工业大学建筑学院

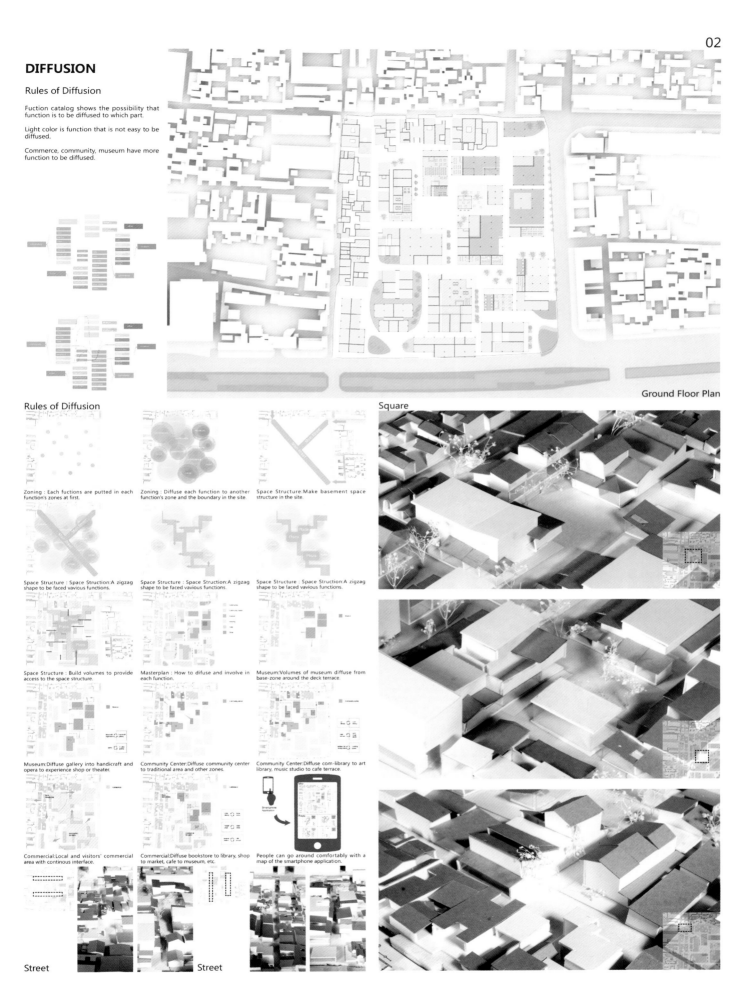

应变，随时而变
基于季节及人流变化的可拆装木结构游客中心

IN WINTER 冬天 | IN SUMMER 夏天 | IN POPULAR SEASON 旺季

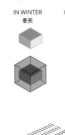

不可变建筑与可变建筑的对比
COMPARISON BETWEEN DEAD AND FLEXIBLE ARCHITECTURE

We can see that an architecture with a flexible system reacts to the environment actively and can meet with different needs. It do not have to change a lot, but will make a big difference in creating a comfortable physical environment.

The design uses natural and local materials (wood)as structural element. It is to create a connection which leaves no impact on the site.

By adding the same joints of wood and connecting them with each other, we get a structural grid.

Based on this grid, we add architecture volume in it according to the required program.

IN POPULAR SEASON 旺季
Remove some columns and beams to make a stage.
旺季去掉一部分建筑形成舞台

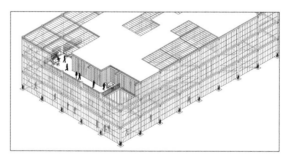

IN WINTER 冬天
Install glass curtain walls to keep warm inside.
冬天装上幕墙保暖

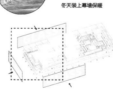
IN SUMMER 夏天
Remove glass curtain walls to allow ventilation.
夏天去掉幕墙通风

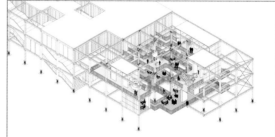

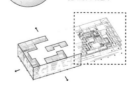

A-A 剖面 1:300

作品名称：应变，随时而变
Change over Time

三等奖

院校名：哈尔滨工业大学
设计人：王疆，李栋梁，钟建博，赵东吉
指导教师：徐洪澎，吴健梅，张纹韶
课程名称：开放式研究型建筑设计
作业完成日期：2015年03月
对外交流对象：英国巴斯大学

应变，随时而变
基于季节及人流变化的可拆装木结构游客中心

总平面 SITE PLAN

The site locates in Dagu mountain holiday resort in Dandong city. It is back on Dagu mountain and near the national road 201. The area is home to long winters, and humid, somewhat hot summers belonging to the humid continental regime.

活动分析

List of Building Construction Layer:
1. Timber Siding
2. Rain Screen Battens(Fixed to 7(Battens))
3. Exterior Sheathing
4. XPS Foam Thermal Insulation
5. Tack Coat
6. Timber Cladding
7. Battens
8. High Density Cellulose
9. Keels
10. Timber Plate
11. Timber Plate
12. Rigid Foam Thermal Insulation
13. Ground Smooth
14. Battens
15. Parquet Flooring
16. Hardwood Cladding
17. Ground Smooth; Separating Layer
18. Vapor Barrier
19. Ground Smooth; XPS Foam Thermal Insulation
20. Tack Coat
21. Cross laminated Timber
22. Plate
23. Keels

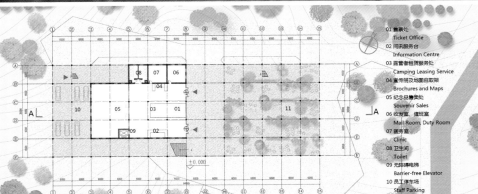

一层平面图 1:400 First Floor Plan

01 售票处 Ticket Office
02 问讯服务台 Information Centre
03 露营者租赁服务处 Camping Leasing Service
04 宣传册及地图自取架 Brochures and Maps
05 纪念品售卖处 Souvenir Sales
06 收发室、值班室 Mail Room, Duty Room
07 医务室 Clinic
08 卫生间 Toilet
09 无障碍电梯 Barrier-free Elevator
10 员工停车场 Staff Parking
11 室外花园及休息场地 Outdoor Garden and Rest

THE WHOLE STRUCTURE CONSISTS OF TWO STRUCTURAL ELEMENTS.

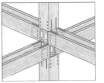 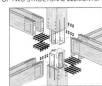

This joint connects the beam and the column. Two ends of the beam and the column are fixed with steel members to make independent structure members. All components are symmetrical, so there is no left or right, up or down. This could avoid mistakes when assembling.

 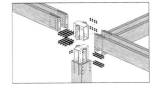

This joint connects the top beam and the column. It is transformed from the first joint for the integrity of the eave.

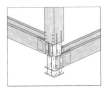 

This joint connects the bottom beam and the column. To avoid corrosion from rain water and moisture, we use steel component to lift the whole structure. It is also transformed from the first joint.

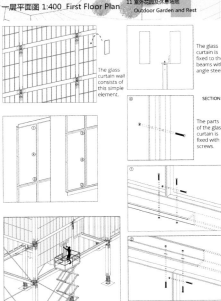

The glass curtain wall consists of this simple element.

The glass curtain is fixed to the beams with angle steel.

SECTION

The parts of the glass curtain is fixed with screws.

THE JOINT OF THE GLASS CURTAIN WALL
The glass curtain acts as a warm envelope in the winter. It is divided into small parts and can be installed easily and quickly by several men in a day. The joint between each element is designed for easy assembling and as light as possible.

University: Harbin Institute of Technology
Designer: Wang Jiang, Li Dongliang, Zhong Jianbo, Zhao Dongji
Tutor: Xu Hongpeng, Wu Jianmei, Zhang Wenshao
Course Name: Harbin Institute of Technology - Bath University Joint Design Studio
Finished Time: Mar., 2015
Exchange Institute: Bath University School of Architecture

Third Prize

平行世界

本课题为四年级中日联合课程设计，由日本京都大学教授和日本三菱地所设计师以及东南大学教师联合讲授。课题结合日本东京、大阪交通综合体的实地考察，让学生对亚洲城市存在的问题有更为深刻的认识，了解从人文关怀出发进行城市设计和建筑设计的思维方法。并以南京拟建的马群地铁站地块为研究对象，探讨如何在人口密集的城市中置入新的交通综合体，以满足城市生活并激发城市活力。

面对当今亚洲城市人口密集、交通拥挤、车水马龙的"世俗"生活，人们只能在远离城市的乡村才能体验桃花源般、坐看云起的"出世"生活。这二者由于它们对场地等客观因素的要求不同仿佛不能在同一时空的同一空间下存在，如同无法相交的平行线。"平行世界"旨在创造一种新的场地模式，在一块场地中同时呈现出两种看似不可能产生交集的生活模式，并试图发现这二者中可能发生的一些有趣的"交集"。

南京地铁马群站城市设计

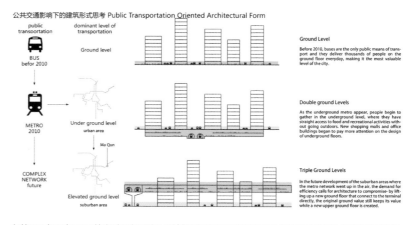

公共交通影响下的建筑形式思考 Public Transportation Oriented Architectural Form

Ground Level
Before 2010, buses are the only public means of transport and they deliver thousands of people on the ground floor everyday, making it the most valuable level of the city.

Double ground Levels
As the underground metro appear, people begin to gather in the underground level, where they have straight access to food and recreational activities without going outdoors. New shopping malls and office buildings began to pay more attention on the design of underground floors.

Triple Ground Levels
In the future development of the suburban areas where the metro network went up in the air, the demand for efficiency calls for architecture to compromise - by lifting up a new ground floor that connect to the terminal directly, the original ground value still keeps its value while a new upper ground floor is created.

Architecture has to keep up with the changing society, and follow the tendency of development. The fast urbanization of Nanjing lead to the spread of its transportation network into suburban areas, and they are the vessel that the city and future architecture have to rely on. By compromising to the future transportation network I actually challenged the traditional acknowledgement of ground level. As gigantic and different as the mega structure may seem, it actually tripled the value of land and allowed us to create a parallel world of purity that stands above our currently chaotic urban life.

交错 Intersactions

The parallel worlds, even though different in their promises of life, are not isolated from each other. Different types of art spaces were embedded to lead people into another world through the exploration of art, which enlightens the tedious urban life with unexpected surprises.

模型照片 Photographs

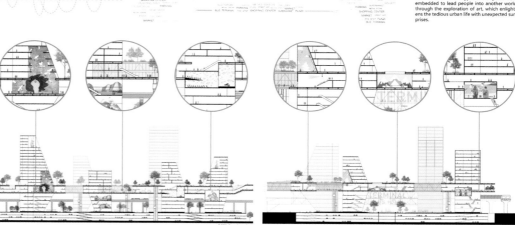

剖面 Section B-B　　剖面 Section A-A

作品名称：南京地铁马群站城市设计
Nanjing Subway Station Urban Design

院校名：东南大学
设计人：张宏宇，罗西
指导教师：唐芃，沈旸，宗本顺三，惠良隆二
课程名称：建筑设计Ⅳ
作业完成日期：2014年11月
对外交流对象：日本京都大学，日本三菱地所

三等奖

PARAWORLD URBAN COMPLEX

This project is the joint teaching program with Kyoto University and Mitsubishi Estate. The visiting of metro stations in Tokyo and Osaka is aimed to enhance the students' understanding of problems of Asian cities and make them build the idea "people-oriented" in urban design. The studio study with the proposed Maqun transfer stations to think about how to intervene a new metro station in a dense city to invitalize the city.

Living in Asia and facing problems of over dense, we create an interlaced world in our urban design, ParaWorld, separating and paralleling the rural and urban parts of our world in one space time. Key joints are connected to become highly centralized, where wonder comes to birth.

The upper ground level
Accessible through elevators and lifts, the upper ground floor is a huge garden that offers variety of outdoor activities and help establish a healthy living environment.

The ground level
Straightly connected to the terminal station, the ground floor is the "living room" of the mega structure. It provide places and activities for people of different purposes to meet, It also organize visual and physical relationship with the other levels.

The conventional ground level
Accessible to all citizens on foot, by bus and by cars, the conventional ground level offers a huge free space under the shelter of the mega-structure. It is a dynamic playground for the visitors and a fancy shopping market that help the local economy to thrive.

剖透视 Section Perspective

轴测分解 Axonometric

I **Buildings** are placed to meet the density requirement of the site.

II **Ground floor** is lifted up on the same height as the metro lines to meet the demand for efficiency, the terminal is included in it.

III **Upper Ground Level** is a peaceful wonderland that stand above the city, it offers an world of purity away from noise and chaos.

IV **Conventional Ground Level** is a place for market and commercal exchanges, with which Parallel World stimulate local business to thrive.

V **Push** the platform to welcome the main crowd from around the site.

VI **Pull up** some parts to get a better view.
Pull down some parts to create access to the ground floor.

VII **Cut** holes in the platform to introduce sunlight and air to the plaza below.

University: Southeast University
Designer: Zhang Hongyu, Luo Xi
Tutor: Tang Peng, Sheng Yang, Munemoto Junzo, Era Ryuji
Course Name: Southeast University Design Studio
Finished Time: Nov., 2014
Exchange Institute: Kyoto University School of Architecture, Mitsubish Estate

Third Prize

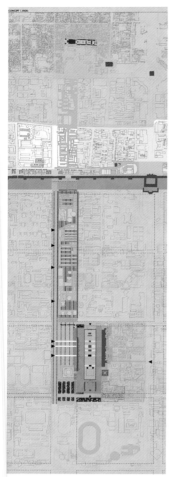
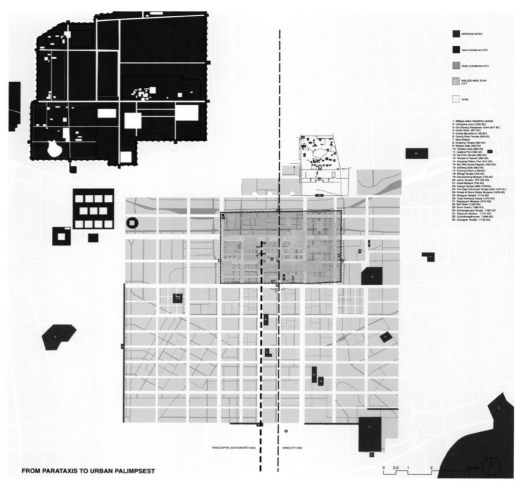
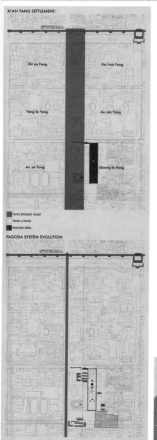

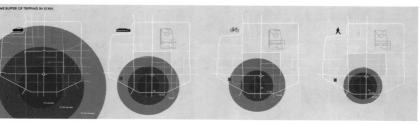

作品名称：传承·规矩——唐轴线小雁塔地段城市设计
Inheritage and Rules-City Design of the Xiaoyan Pagoda Area in the Axis of Tang Dynasty

院校名：西安建筑科技大学
设计人：Michele Marini, Claudio livetti, Daniale Delgrosso, 廖枢丹, 孙雅雯, 邢泽坤
指导教师：李昊, 常海青, 鲁旭, 李煜, laura Anna Pezzetti, Carlo Palazzolov
课程名称：2015西安建筑科技大学—意大利米兰理工大学国际联合工作营课程教学
作业完成日期：2015年05月
对外交流对象：意大利米兰理工大学建筑学院

三等奖

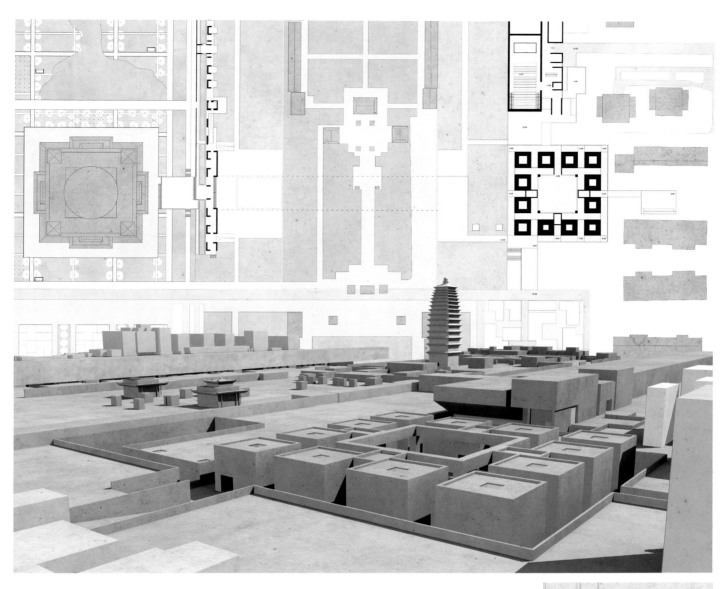

University: Xi'an University of Architecture and Technology
Designer: Daniele Delgrosso, Claudio Livetti, Michele Marini, Liao Shudan, Xing Zekun, Sun Yawen, Xin Zhekun
Tutor: Li Hao, Chang Haiqing, Lu Xu, Li Kun, Laura Anna Pezzetti, Carlo Palazzolo
Course Name: Worshop Politecnico di Milano, Xi'an university of Architecture and Technology
Finished Time: May, 2015
Exchange Institute: Politecnico di Milano, Xi'an University of Architecture and Technology

Third Prize

TANG AXIS

WALKING ON SILK ROAD
A CITY BASED ON A MODE OF HERITAGE PROTECTION IN RESIDENTIAL

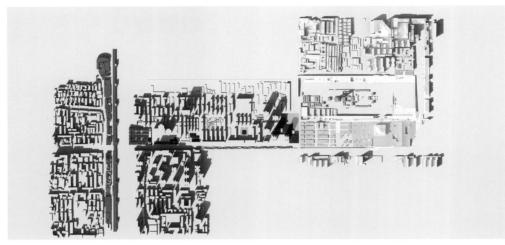

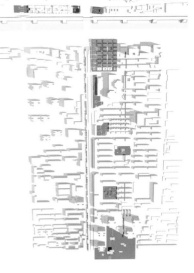

The Tang axis public space is connected by four seasons park, they are spring park, summer park, autumn park and winter park Respectively.We take the residential district as the basic unit for the protection of ancient sites, and improve the quality of life of the surrounding residential area. The four seasons park is reduction the Zhuque Avenue scene at the time of the Tang Dynasty that described by the poet, and fuse the development of the Silk Road.The park connect the small wild goose pagoda as a starting point, end up with the city wall, the design is making people walk on the Silk Road with a different attitude again.

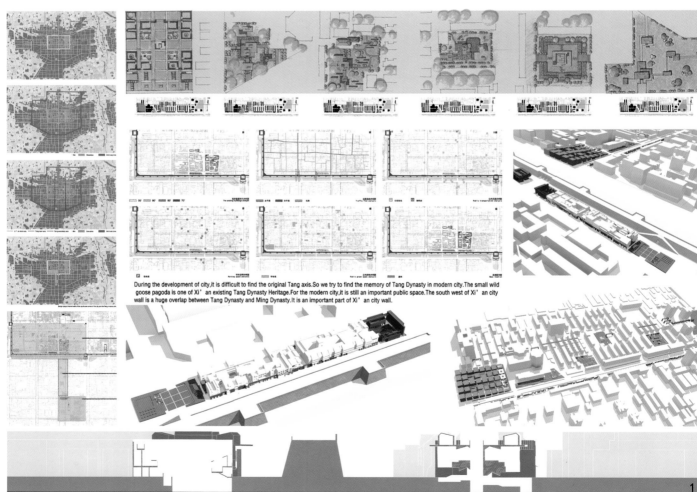

During the development of city,it is difficult to find the original Tang axis.So we try to find the memory of Tang Dynasty in modern city.The small wild goose pagoda is one of Xi'an existing Tang Dynasty Heritage.For the modern city,it is still an important public space.The south west of Xi'an city wall is a huge overlap between Tang Dynasty and Ming Dynasty.It is an important part of Xi'an city wall.

作品名称：行走在丝路之上
Walking on the Silk Road

院校名：西安建筑科技大学
设计人：尹锐莹，Alberto Malabarba，Francesco Busnelli，路冠丞，张雅楠，张婧琪
指导教师：李昊，常海青，鲁旭，李煜，Laura Anna Pezzetti, Carlo Palazzolov
课程名称：2015西安建筑科技大学—意大利米兰理工大学国际联合工作营课程教学
作业完成日期：2015年04月
对外交流对象：意大利米兰理工大学建筑学院

三等奖

STRATEGY

site

courtyard

vertical

improve the courtyard
contact surrounding space

DESIGN SPECIFICATION

The Silk Road started from the ancient China and connected with the ancient road commercial trade routes between Asia and Europe. In the Western Han Dynasty, the first time Zhang Qian opened up the Silk Road, in the Eastern Han Dynasty, Ban Chao operated in the western regions and through extension of the Silk Road again. Silk road opened up contribution greatly to the eastern and western economic, cultural, religious, linguistic intercommunication and fusion.
The design aims to arouse memories and feelings of the ancient city. In contemporary urban, we can feel its presence, just like we walk on the Silk Road once again. In the Tang axis, we selected some node space to design, like the Silk Road set with some pearls. Through the five senses design, people can experience the glory of Tang Changan again.
We can see the morden urban space enriched by the superposition over time, formated the multiple space city, merged into the two points of city. The future development of the city is still on the way, we should continue to shine the bright spot.

WALKING ON SILK ROAD
A CITY BASED ON A MODE OF HERITAGE PROTECTION IN RESIDENTIAL

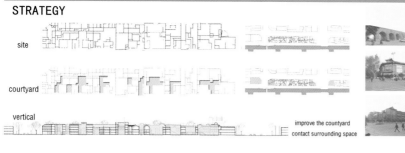

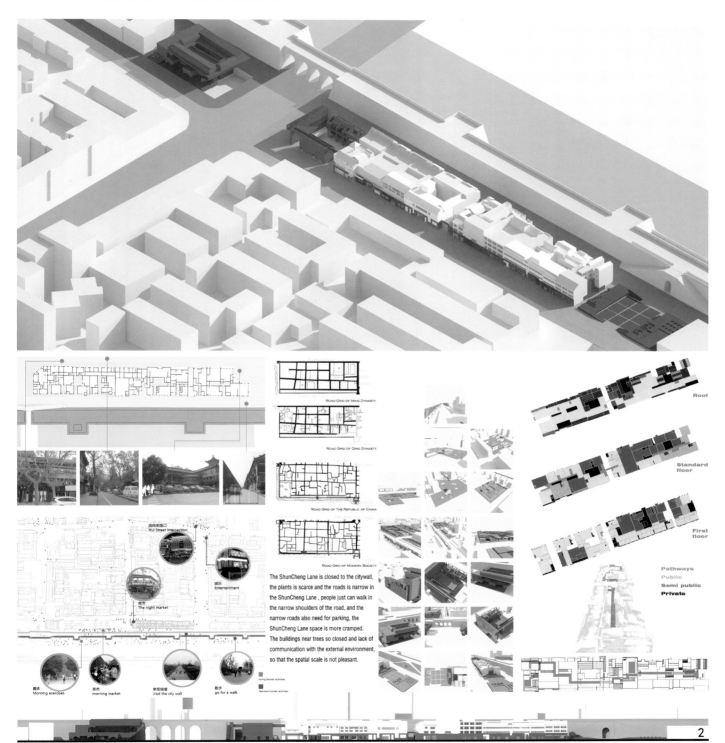

The ShunCheng Lane is closed to the citywall, the plants is scarce and the roads is narrow in the ShunCheng Lane, people just can walk in the narrow shoulders of the road, and the narrow roads also need for parking, the ShunCheng Lane space is more cramped. The buildings near trees so closed and lack of communication with the external environment, so that the spatial scale is not pleasant.

University: Xi'an University of Architecture and Technology
Designer: YinRuiying, Alberto Malabarba, Francesco Busnelli, Lu Guancheng, Zhang Yanan, Zhang Jingqi
Tutor: Li Hao, Chang Haiqing, Lu Xu, Laura Anna Pezzetti, Carlo Palazzolo
Course Name: Xi'an University of Architecture and Technology - Politecnico di Milano Workshop City Wall Lane Area
Finished Time: Apr., 2015
Exchange Institute: School of Architecture, Politecnico di Milano, Italy

Third Prize

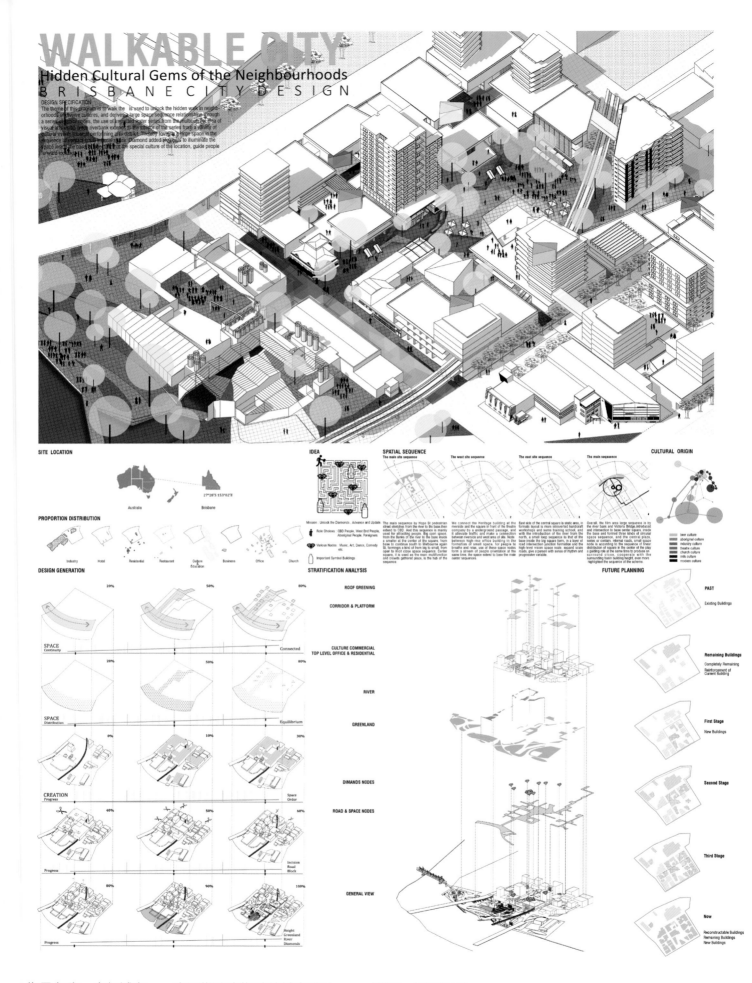

作品名称：步行城市——布里斯班韦斯滕德城市设计
Walkable City

三等奖

院校名称：山东建筑大学
设计人：朱轩毅，贾慧，殷子君，党常硕
指导教师：张克强，任震，Yvonne Wang，Tony Van Raat
课程名称：山东建筑大学—澳大利亚昆士兰理工大学联合设计课程教学
作业完成日期：2015年6月
对外交流对象：澳大利亚昆士兰理工大学

WALKABLE CITY
Hidden Cultural Gems of the Neighbourhoods
BRISBANE CITY DESIGN

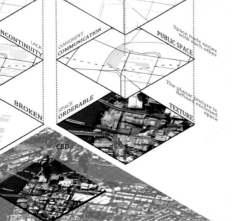

University: Shandong Jianzhu University
Designer: Zhu Xuanyi, Jia Hui, Yin Zijun, Dang Changshuo
Tutor: Zhang Keqiang, Ren Zhen, Vonne Wang, Tony Van Raat
Course Name: Shandong Jianzhu University-Queensland University of Technology Joint Design Studio
Finished Time: Jun., 2015
Exchange Institute: Queensland University of Technology

Third Prize

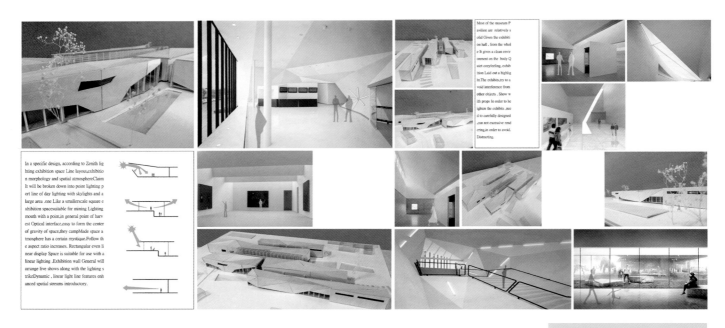

普拉托历史博物馆设计
PRATO MUSEUM DESIGN

GAP · LIGHT 壹

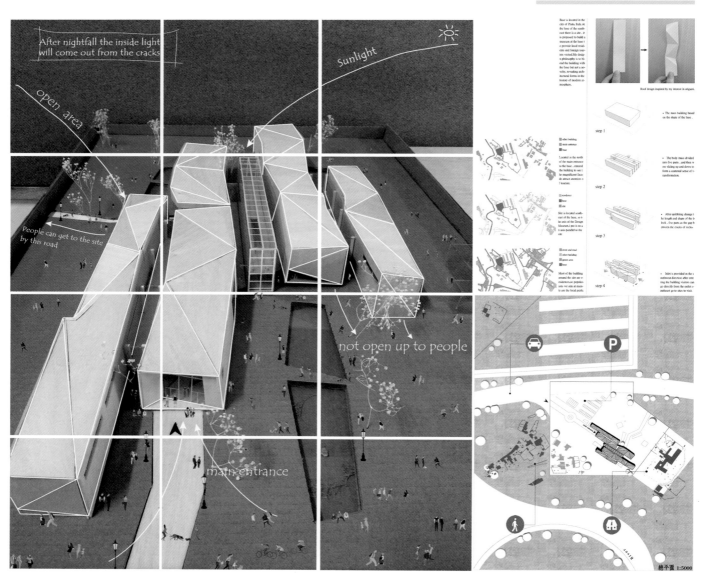

作品名称：隙光——意大利普拉托考古遗址博物馆设计
Gap and Light: Prato Museum Design

院校名：山东建筑大学
设计人：仲文
指导教师：慕启鹏，孔亚暐，Tony Burge
课程名称：建筑设计3
作业完成日期：2014年11月
对外交流对象：新西兰Unitec理工学院

三等奖

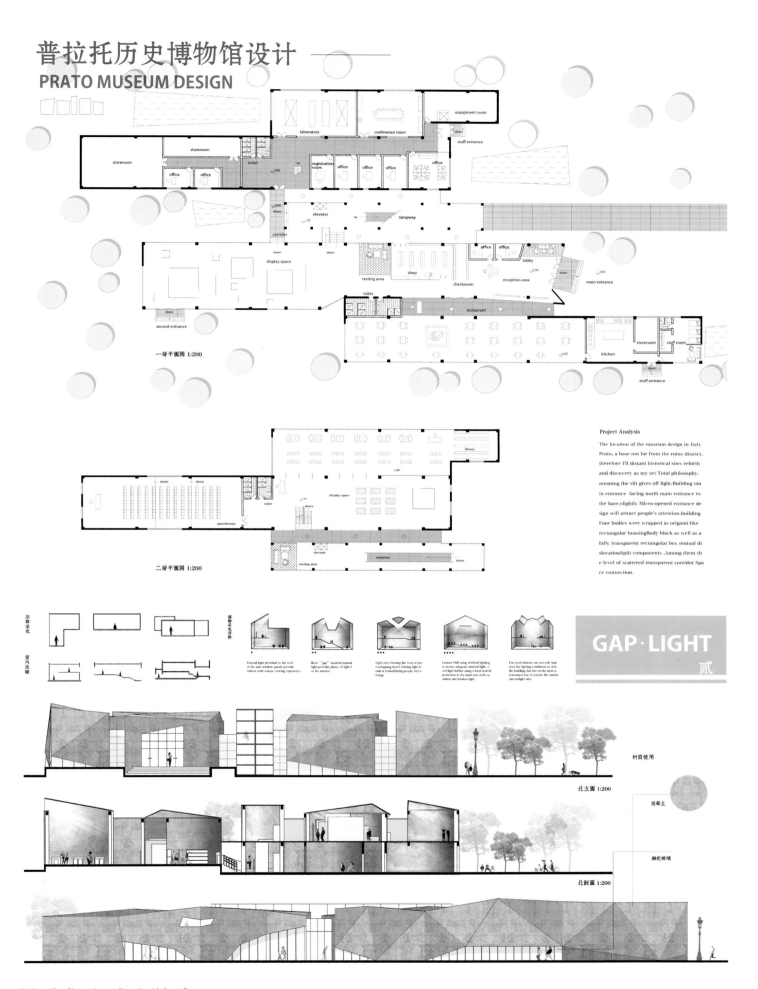

越陌度阡
WALKING THROUGH THE STREETS

太原市五一路及精营历史街区更新计划
Design of public space in the five one road and Jing Ying area of Taiyuan City

设计说明 Design explanation

随着城市的快速发展，暴露出越来越多的城市问题，传统与现代的矛盾突出；道路交通体系充分凸显，传统的街巷体系遭到破坏，新的城市道路交通缺乏活力。通过空间量化分析，研究该区域的道路特性，从而对症下药。

太原市五一路及精营历史街区属于太原府城，具有悠久的历史。街巷纵横，通过一系列局部的改造，逐步梳理出一种新的道路体系，人们越陌度阡，在不同的时空之间穿梭。

With the rapid development of the city, more and more urban issues have been exposed, especially the problem of transport system, the destruction of the traditional street system, and the new urban road lack of energy. How to coordinate the relationship of the tradition and modern? By quantifying spatial analysis and researching road characteristics of the region we can resolve these problems.

Taiyuan Wuyi Road and its historic district which belong to the old part of Taiyuan has a long history with many streets crossing in it. Through a series partial transformation, we can gradually tease out a new road system. When people walk through the footpath, they will feel just like walking through different spaces of different time.

计算分析 Calculation
计算步骤 Calculating step

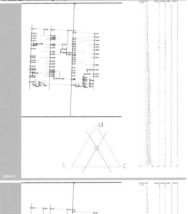
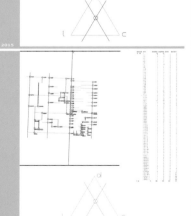

府城变迁 Changes in the ancient city

 宋 (960-1127) Song (960-1127)
宋太原城距今近九百年。据《阳曲县志》所载："宋之太平兴国七年建，俗于四南，明洪武九年，永平侯晋成展东、南、北三面"，宋太原城在明城的西南角。
Song Taiyuan city dating back nearly nine hundred years. Song Taiyuan city in the southwest corner of the city of the Ming.

 明 (1368-1644) Ming (1368-1644)
明太原城在正南北的方形城市，它是一座大三重环套者，明洪武三年，南华门历史街区属于晋王府的宫城范围，在府城中占据了主要的优势地位。
Taiyuan City of north-south out of the city square, Hongwu years, SouthGate.Historic District Jin palace belonging to the scope of Miyagi, occupy a prominent position in the city with the advantage.

 清 (1616-1911) Qing Dynasty (1616-1911)
清康熙正中年，南华门历史街区属于清军的驻营范围，是府城中特殊管辖的区域名。
Qing Yongzheng, South Gate Historic District belong to the scope of the Qing barracks, regional city with special controls.

 民国 (1912-1949) Republic of China (1912-1949)
同期国时期，南华门历史街区集中了大量的名人故居，是府城中的"高端"居住区。
Republic of China, South Gate historic district a large concentration of dignitaries, Tainan is in the "high end" residential area.

NOW
建国初行政中心
The administrative center of the early founding
建国初期，南华门一带曾一度作为新政府的政治活动中心。
The founding SOUTH Mun once as the new government's political center.

道路演变 Road evolution

2002 / 2007 / 2012 / 2015 / FUTURE

2002	绝对值	相对值
连续性 (l)	208	0.26
连接性 (c)	227	0.29
深度 (d)	353	0.45
总值 (s)	788	1

计算说明：通过2002年、2015年和方案的道路统计分析，可以看出交通的连续性以及连接性均有所提高，而深度值下降，从而说明方案的有效性。连续性与连接性的提高最直接体现在住民的通勤，连接当地居民交流的加强便捷，深度的降低使得内外空间联系便捷。

Calculation Description: By 2002, the Road Statistics 2015 and program analysis, we can see the traffic continuity and connectivity are improved, and the depth value decreases, thereby illustrate the effectiveness of the program.

2015	绝对值	相对值
连续性 (l)	134	0.28
连接性 (c)	146	0.3
深度 (d)	205	0.42
总值 (s)	485	1

future	绝对值	相对值
连续性 (l)	169	0.28
连接性 (c)	196	0.32
深度 (d)	248	0.4
总值 (s)	613	1

总平面图 Site plan

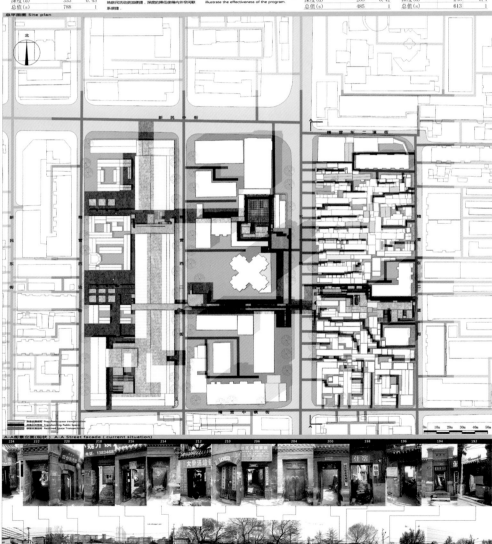

A-A街景立面(现状) A-A Street facade (current situation)

作品名称：越陌度阡
Walking through the Streets

三等奖

院校名：太原理工大学
设计人：郝志伟，罗文婧
指导教师：徐强，高静，董艳平，Karel Nieuwland
课程名称：历史街区更新工作坊
作业完成日期：2015年06月
对外交流对象：荷兰Karel事务所

越陌度阡
WALKING THROUGH THE STREETS

太原市五一路及精营历史街区更新计划
Design of public space in the five one road and Jing Ying area of Taiyuan City

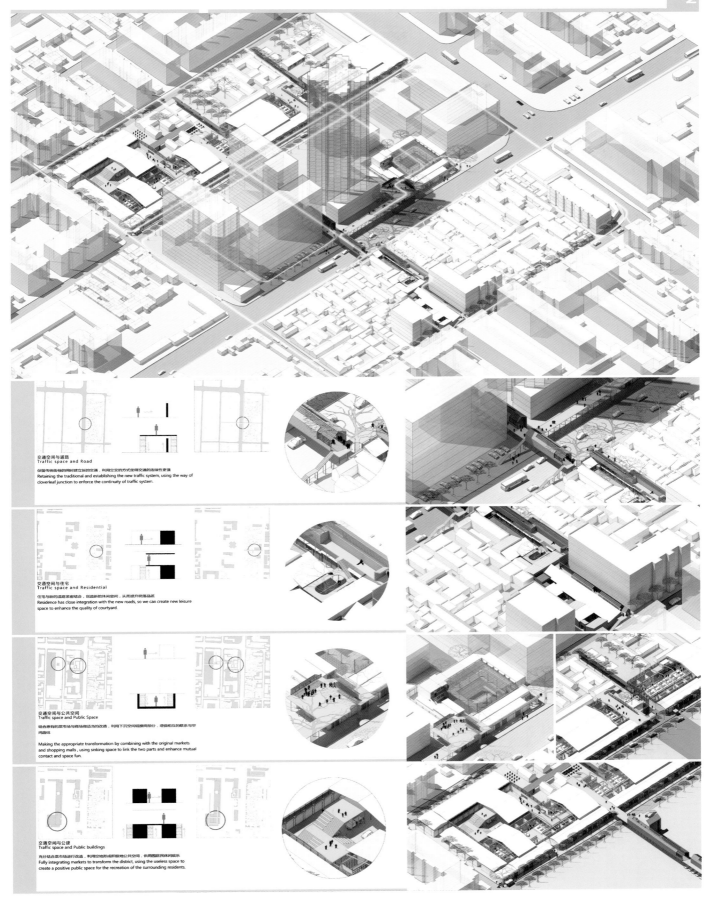

University: Taiyuan University of Technology
Designer: Hao zhiwei, Luo wenjing
Tutor: Xu qiang, Gao jing, Dong yanping, Karel
Course Name: Taiyuan university of technology-Karel Nieuwland Architekten-Redevelopment of historic district of workshop
Finished Time: Jun., 2015
Exchange Institute: Karel Nieuwland Architekten

Third Prize

CAMBRIDGE
colleges
departments

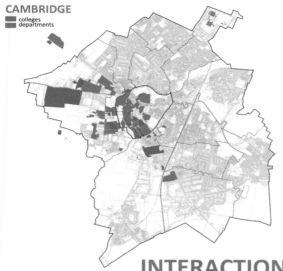

CITY CENTRE

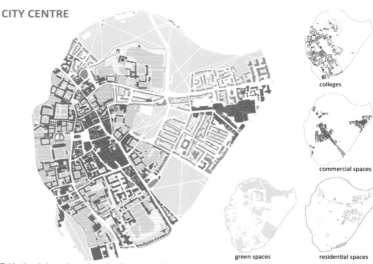

colleges
commercial spaces
green spaces
residential spaces

INTERACTION
ANALYSIS OF COMMON SPACES IN THE COLLEGES IN THE UNIVERSITY OF CAMBRIDGE

Thriving through almost a thousand year, the college system has been the most distinctive feature for the university of Cambridge. With a highly hierarchical and complete structure, the college serves as a self-governing corporation to manage their members and affairs with good facilities and environment for social and academic life. Supported by the circumstance, members of the college are strongly connected by the college which constructs a shared belonging and affinity between the members from now to the future. However, this system is so exclusive to the town in terms of the physical and psychological prohibition and inaccessibility. It is evident that members of the university easily construct a shared and collective identity whilst town people are put in a subordinate position and establish a outsider identity to the university and even to the city. To break the division of town and gown and the feeling of insider and outsider, a shared community should be built up to enhance the relationship and collaboration.

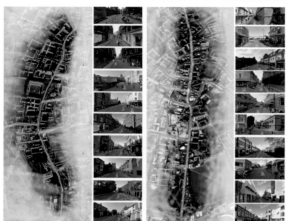

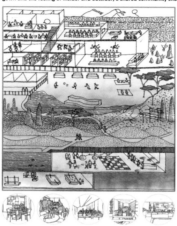

THE INTERFACE OF THE CITY
Walking in the main streets of Cambridge, you would find out that the colleges, exclusive for town people while inclusive for gown people, dominate the interface of the city. A warning note has been attached in Christ College: 'Only college members and visitors hosted by the member are welcomed.' The residents are ruled out of the college communities which play such an important part in the city.

UTOPIA
Each college has its own common room, dinning hall, chapel, library and private rooms, by which the members can easily establish their shared identity and collective belonging in this strong community.

MECHANISM
With an enclosed and exclusive mechanism, the collegiate system obviously establishes a hierarchy and division between the insider and outsider, the privileged and unprivileged, accounting for the hostility and antagonism between them.

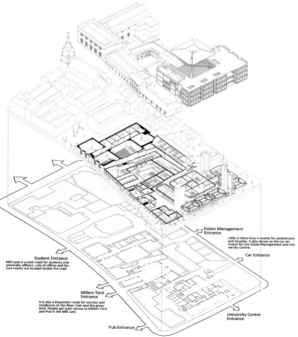

Estate Management Entrance

Little st Mary lane is mainly for pedestrians and bicycles. It also serves as the car entrance for the Estate Management and University Centre.

Student Entrance
Mill Lane is a main road for students and university officers. Lots of offices and lecture rooms are located beside the road.

Car Entrance

Millers Yard Entrance
It is also an important route for tourists and residences to the River Cam and the grassland. People get main access to Millers Yard and Pub in the Mill Lane.

University Centre Entrance

Pub Entrance

SITE ANALYSIS
The activities carried out within the Mill Lane have changed so much considering the last 500 years. But one common has existed in this area that it has been a shared space within different groups. Before, the area was an important commercial area dominated by mills, local trades and inns. After the expansion of the press and the university teaching space, the industrial uses ceased and were developed for university. As the Millers Yard has been purchased by the colleges, this town area would disappear with a redeveloping project as a dormitory. Considering that the University centre has been of poor use, it is a good experimental site to examinate the project in terms of the profits of town and gown.

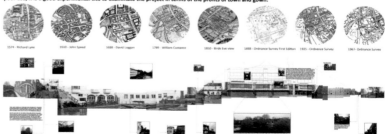

作品名称：互动——剑桥大学学院制下公共空间的分析与设计
Interaction : Analysis of Common Spaces in the Colleges in the University of Cambridge

院校名：南京大学
设计人：黎乐源
指导教师：窦平平，Ingrid Schröder
课程名称：南京大学—剑桥大学联合毕业设计
作业完成日期：2015年06月
对外交流对象：英国剑桥大学建筑系

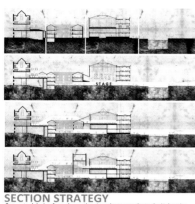
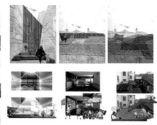
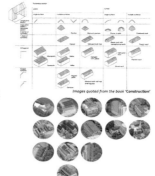
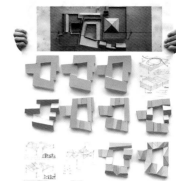

SECTION STRATEGY
Conceptual design from the section. In order to reactivate the University Centre and rennovate the Miller Yard, the design is inspired from the way of demolishing anything to the way of making use of the existed buildings.

RENNOVATION STRATEGY
Conceptual design from parts of the site. Explore the possibilities of rennovation in the existing buildings.

ROOF ANALYSIS
Analyze the roof context in Cambridge.

TRANSFORMATION
The transformation of the roof is determined by the roof context and the space undereath.

CIRCULATION GALLERY
With a strategy of connecting the Lecture Hall and The University and a strategy of attracting people to the rooftop by the tempting facade, we can make full use of the rooftop and transform it into an art centre to set up the mechanism of circulation with an attempt to achieve the interaction between town and gown.

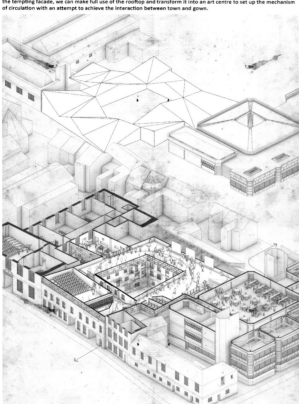

University: Nanjing University
Designer: Li Leyuan
Tutor: Dou Pingping, Ingrid Schröder
Course: Cambridge Design Research Studio
Finished Time: Jun., 2015
Exchange Institute: Department of Architecture and Historical Art, University of Cambridge

Third Prize

KALEIDO CITY 01 //Eco-cities Urban Design Studio 2015

SKELETON MODEL Scale 1/200

In order to get more effective benefits from the flow of nature and reduce pollution, the concept of "Kaleido City" is being investigated. Using some principles from traditional civilian wisdom for reference, studies of how the original residents created a livable and sustainable settlement without high technology or huge resource input are being done. The research showed that making the maximal use of original resources is the best way to achieve this purpose. In addition, some natural and human resources at the site could be combined in a coexisting circulation loop, catering to these ecological ideas.

PROTOTYPE

In Chinese philosophy, the famous concept "FengShui" means understanding nature well then utilizing it in a better way. When looking at the cracks on charred wood, we found that the bark was divided into blocks in a form where there is order in chaos. The texture depends on various parameters, such as material density, surface curvature and burning temperature. Urban morphologies might be generated by many natural factors similarly. Inspired by that, we extract the organic settlement patterns from the bark to be the design prototype, and then define the building heights, orientations, spatial sizes and street directions scientifically in terms of local sunlight and wind analyses.

MATERIALS & PATTERNS

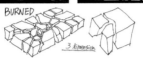

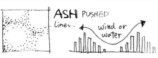

POTENTIALS

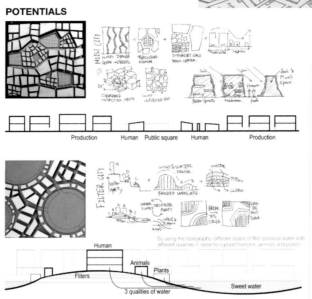

By using the topography, different layers of filter produce water with different qualities in order to support humans, animals and plants.

SUNLIGHT & WINDS

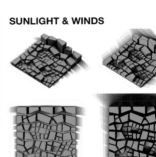

Summer Solstice   *Winter Solstice*

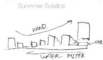

Referring to the sunlight simulation and wind analysis, we turned the buildings orientation to a North-South facing and adjusted their heights in order to get more sunshine in winter.

TARGET GROUP

We focus on the group of children and seniors, who stay in the city much longer than the adults who go far away to work everyday. "Kaleido City" has three compact centers all serving public purposes. The middle one with a plaza is an educational building group offering schools and libraries, while the other two in the southern part, which receive great sunlight, are senior centers and rain-collecting fish ponds.

FUNCTIONAL MODEL Scale 1/500

AGRI-SURFACES

FARMING LAND
Rice & Bamboo

ROOF GREENHOUSE
Fruit & Vegetables

ROOF FARMLAND
Root & Stem Plants

SOUTH AGRI-WALL
Dwarf Local Plants

EAST & WEST WALL
Creepers & Windows

FUNCTIONS

RESIDENT

SENIOR HEALTH CENTER

SCHOOL

KINDERGARTEN

LIBRARY & CENTER

SPORTS CENTER

MARKET

STORE

PLAZA

FISH POND

作品名称：万花城——人与生态的城市设计
Kaleido City

院校名：浙江大学
设计人：严子君
指导教师：Joerg Baumeister, Daniela A. Ottmann, 吴越, 王卡
课程名称：专题化设计
作业完成日期：2015年06月
对外交流对象：西澳大利亚大学，澳大利亚城市设计研究中心

三等奖

KALEIDO CITY 02

ECOLOGICAL SYSTEM

The above diagram shows the correlation of the natural system (green) and human system (blue). The main idea of an eco-system is maximizing outputs as well as reducing pollution and waste. With a concern for water recycling, four streams have been designed in the site. They are running from north to south, collecting the waste water from each family, and then inflowing to the fish ponds after being purified by precipitation and microbes. A big system is constituted by many smaller systems, such as the bamboo-fishpond coexisting loop, which reuses the skin from bamboo shoots as fish food. Furthermore, the streets are designed in a smaller scale and organic shapes are used to create closer connections in the neighborhood and to better explore the possibilities of experience for pedestrians.

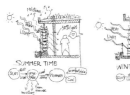

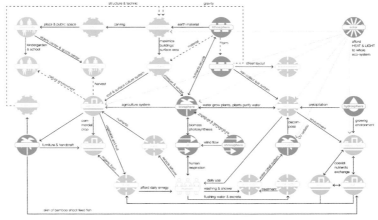

WATER RECYCLING FLOW

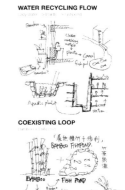

COEXISTING LOOP

ECO STRATEGY

The combination of neglected spaces and human resources can produce great extra output. Generally, the productive potentials of buildings exterior walls, neglected in cities, are able to receive natural resources such as solar energy and rain directly. The strategy of this project is to use low-rise high-density dwellings to maximize the city surface, in an attempt to improve productivity. Through reasonable arrangement, about 20,000 square meters of productive surfaces (south facades and roofs), receiving excellent sunshine all year round, would be created in the site. It can supply half of the vegetable consumption for 800 residents annually with column-type soil free cultivation and roof-garden planting. This self-satisfying model also provides learning chances for children and activities for seniors when they are spending...

POPULATION

CALCULATION - PRODUCTIVITY

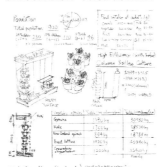

ENVIRONMENTAL COMPATIBILITY

The most valuable features of "Kaleido City" are its great environmental compatibility and flexibility. With systems in different scales, such as a coherent roof-planting network, rainwater outlet and recycle system, a bamboo-fishpond coexisting loop, and highly flexible facade system, "Kaleido City" can transform and adapt to different climates, site shapes, populations and urban functions — good for further application and suburbanization.

SAME SURFACE, VARIOUS AGRICULTURAL STRUCTURES

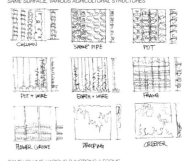

SAME VOLUME, VARIOUS FUNCTIONS & FORMS

ARCHITECTURAL DRAWINGS

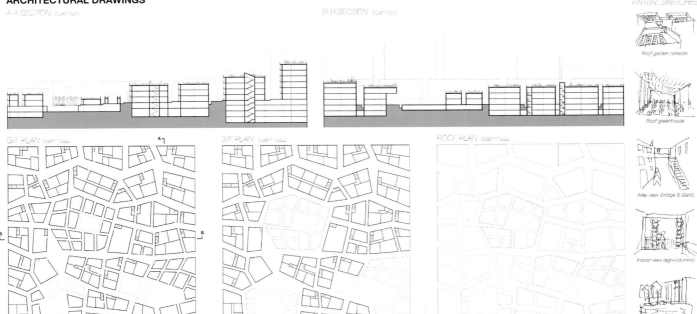

A-A SECTION   *B-B SECTION*   *PARTIAL SKETCHES*

G/F PLAN   *3/F PLAN*   *ROOF PLAN*

Roof garden network
Roof greenhouse
Alley view (bridge & stairs)
Indoor view (agri-columns)
Fishpond and bamboo

University: Zhejiang University
Designer: Im Chi Kuan
Tutor: Joerg Baumeister, Daniela A. Ottmann, Wu Yue Wang Ka
Course: Zhejiang University - The University of Western Australia - Specialized Topics in Architecture Design
Finished Time: Jun., 2015
Exchange Institute: The University of Western Australia, Australian Urban Design Research Centre

Third Prize

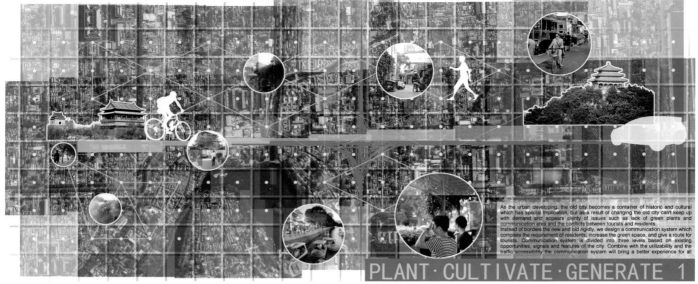

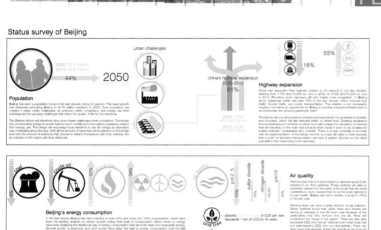

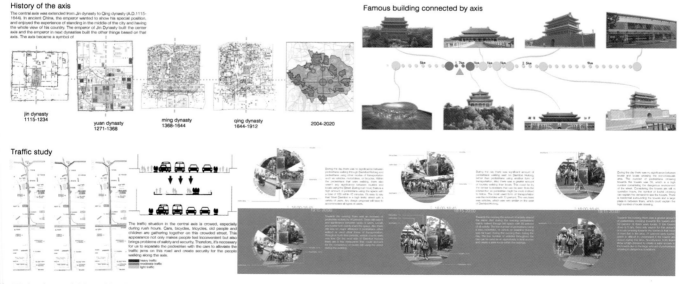

作品名称：种植・培养・建造
Plant・Cultivate・Generate

院校名：北方工业大学
设计人：张雅琪，林悦，王悠然，Cyndi Feris, Mutsawashe Chipfumbu, Umurinzi Serge, Niyoyita Joseline, Zenas Guo, Alan Hu, Lorena Jauregui, Artuno Ortuno, Julianne Pineda, Eddy Solis, Joseph Jamoralin, Hakizimana Dusenga Yvette
指导教师：秦柯，Irma Ramirez, Courtney Knapp, 张勃，安平
课程名称：北方工业大学—加利福尼亚州立理工大学伯莫纳分校联合设计（本科四、五年级）
作业完成日期：2015年07月
对外交流对象：美国加利福尼亚州立理工大学伯莫纳分校

PLANT · CULTIVATE · GENERATE 2

Traffic accessibility

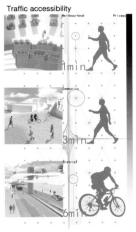

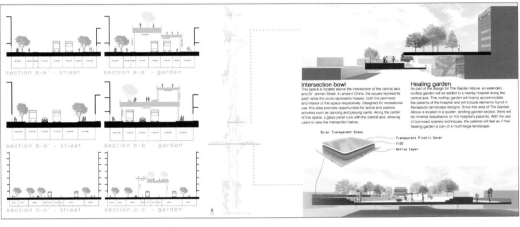

Intersection bowl
This space is located above the intersection of the central axis and Di' anmen Street. In ancient China, the square represents earth while the circle represents heaven, both the perimeter and interior of this space respectively. Designed for recreational use, this area provides opportunities for active and passive activities such as dancing and playing cards. Along the center of this space, a glass panel runs with the central axis, allowing users to view the intersection below.

Healing garden
As part of the Design for The Garden Above, an extended rooftop garden will be added to a nearby hospital along the central axis. This rooftop garden will mainly accommodate the patients of the hospital and will include elements found in therapeutic landscape designs. Since this area of The Garden Above is located in a quieter, strolling garden section, there will be minimal disturbance on the hospital's patients. With the use of borrowed scenery techniques, the patients will feel as if their healing garden is part of a much larger landscape.

Pocket space
Pocket space is the basic level of the organic communication system. They are located in anywhere of the Hutong with an available opportunity, such as half-surrounding areas, ancient trees, old stages, yards and so on. Based on the Hutong 's features and the people gathered there, the pocket space is given a theme which serviced for a group of people, in order to influence the residents in a positive way.

Modality of plants

Plants on cables
By connecting cables from one rooftop to another, the cables acted as a new space for vines to grow. This connection not only added green elements to the Hutong, it also provided a canopy for the people walking below.

Traditional hutong
Because of the original Hutong design, residents were forced to plant within the courtyard because the streets did not accommodate for green space. Today the original trees planted arched over to provide shade both within the courtyard and to the adjacent street.

Wall planter boxes
Within the newer developments, wooden planter boxes were attached to the wall to allow for planting space without taking up any walking space.

Green roof
Typically seen on top of residential high-rises within cities, green roofs treat rooftops as garden space. Again, residents were taking advantage of spaces to plant for themselves.

Roadside planters
Newly implemented in the Hutongs, these planters are a way to add vegetation without breaking into the ground. However, many of these were too bulky and took up too much of the sidewalk, forcing pedestrians into vehicular traffic.

Green facade
By growing plants along the walls, residents and business owners were able to create a lot of green space without taking up any street space.

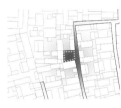

1. CHildren spaces
This pocket space based on a half-surrounding area in which built a dismountable facility to provide a safety and interesting area for children.

2. Entrance spaces
We focused on an old tree grows in the access of a Hutong which connecting with several families. And improve the experience staying here through add facilities and outdoors furnitures.

3. Opera spaces
Residents in Hutong areas love Peking Oprea very much. They are both actors and audiences. So opera spaces are built in a comparative commodious area bring a stage for residents to show themselves.

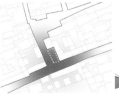

4. Activitiy spaces
We choose yards as a prototype of activity spaces, because they are a significant part in residents daily life for sports. And we make it not only a sports area but also a communication area.

University: North China University of Technology
Designer: Zhang Yaqi, Lin Yue, Wang Youran, Cyndi Feris, Mutsawashe Chipfumbu, Umurinzi Serge, Niyoyita Joseline, Zenas Guo, Alan Hu, Lorena Jauregui, Arturo Ortuno, Julianne Pineda, Eddy Solis, Joseph Jamoralin, Hakizimana Dusenga Yvette
Tutor: Qin Ke, Irma Ramirez, Courtney Knapp, Zhang Bo, An Ping
Course Name: NCUT - Cal Poly College of Environment Joint Design Studio (from grade 4 to grade 5)
Finished Time: Jul., 2015
Exchange Institute: Cal Poly College of Environment

Honorable Mentian

Methodology
Our site analysis work is split up into three sections: Understanding the Past, Understanding Current Rituals, and finally, Envisioning the Future. Our methodology for collecting data was a mixture of internet research and site visits in order to gather knowledge about the space and how community members react to it.

History
After careful research, there were several significant historical events in which was beloved had a strong impact on development in the district. All the events on this timeline drastically changed the social, cultural and physical shape of the area. In order to plan for the future, it is imperative to have a strong understanding of how the current situation came to be.

2008
Beijing Olympics
One of China's motives for holding the Olympics in Beijing was to display the new image of Made-in-China to the world. Before 2008, the Chinese government set forth with new construction and redevelopment goals to update the dated infrastructure which has persuaded locals in moving out of the area.

Today
Present Day
After Olympic era, the government realized the importance of avoiding brand name architecture. In October of 2014, the President of China called for "no more weird buildings" to be built, particularly in Beijing. This opened the gate to smarter urban design efforts.

Bicycle Evolution
The rise of the bicycle in China parallels our other historic events that influenced China to expand on their bicycle infrastructure. Understanding how the population acclimates to new technologies is important as we are trying to implement new ideas. In our project site it is apparent that existing infrastructure is not enough to meet the needs of the modern human. Bicycles are in a similar condition where they are heavily utilized in a space that does not support them.

Initial Entry and Slow Growth: 1900s to 1978
In the early 1990s, the use of bicycle transportation was initially a luxury in China and was only utilized by rich people. During the first two phases, the government promoted bicycle ownership by relating it to economic growth. The number of bicycle ownership slowly increased over time.

Rapid Growth: 1978 to 1995
The economic reform allowed for a wider social class to afford bikes. Bike lanes and other infrastructure accommodations were built during this time. As the number of bicycle ownership increased, China was characterized as a "Bicycle Kingdom."

Bicycle Use Reduction: 1995 to 2002
Two things led to the reduction of bicycles. Government opinions changed during the mid-1990s as the high number of bicyclists on the road began to create traffic problems. Additionally the rise of the electric bike and automobile.

Policies Diversification: 2002 to present day
Government opinions shift once again in 2002, when bike ownership is seen as a way to alleviate congestion. In 2009, the Government published an Action Plan. According to this plan, Beijing will have 1,000 public bike sharing and bike rental stations, with at least 50,000 bikes available near major rail stations and bus terminals. by the end of 2012.

Today
Much like housing and preservation perspectives have changed, it is safe to assume that bicycle infrastructure will also reflect this trend. China is adopting smart solutions to congestion problems.

Social Makeup
After conducting interviews and further examining the project site, three main social groups were identified that encompass Shichahai. These three groups consist of Locals, Tourist and Workers. Each specific group was identified and categorized by their appearance. Locals were identified as anyone passing through the surveyed area, showing no interest to the scenery. Tourist were identified as anyone carrying a camera, examining their surroundings as they walked through surveyed area, or any other features that may give the impression that they are a visitor. Workers were identified as anyone wearing a uniform or carrying an item that may be used for work.

workers   locals   tourists

Afternoon Counts		Night Counts	
Locals	84%	Locals	52%
Tourist	12%	Tourist	42%
Workers	4%	Workers	8%
Male	51%	Male	65%
Female	49%	Female	35%

*288 people counted

Afternoon Counts		Night Counts	
Locals	46%	Locals	52%
Tourist	37%	Tourist	42%
Workers	17%	Workers	6%
Male	38%	Male	65%
Female	62%	Female	35%

*448 people counted   *736 people counted

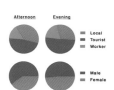

Afternoon Counts		Night Counts	
Locals	51%	Locals	98%
Tourist	44%	Tourist	2%
Workers	5%	Workers	0%
Male	54%	Male	29%
Female	46%	Female	71%

*448 people counted   *268 people counted

钟楼湾胡同
大街

作品名称：地球村
The Global Village

入围奖

院校名：北方工业大学
设计人：薛皓硕，陈婉钰，苏伊莎，艾克拜尔，Uwimana Lydia, Nurlan Babayev, Tinashe Justin Machamire, Emily Williams, Houra Khani, Sergio Gutierre, Juan Galvan, Renzo Pali, Pan Chunguang, Ihumere Irma, Mukazera Marie Christelle
指导教师：秦柯，Irma Ramirez, Courtney Knapp, 张勃，安平
课程名称：北方工业大学—加利福尼亚州立理工大学伯莫纳分校联合设计（本科四、五年级）
作业完成日期：2015 年 07 月
对外交流对象：美国加利福尼亚州立理工大学伯莫纳分校

THE GLOBAL VILLAGE

Improving residents quality of life is our main goal. Our approach for this design is to create better living conditions with the usage of simple modular boxes. These boxes are four different sizes and are used to create functional yet compact spaces for the residents. Based on the activity in the space, the opening and flexibility of the module unit changes. This arrangement of courtyard clusters become the heart and center of the community. The Educational Village is made up of five clusters, which include the historic, technology, self-sustenance, cultural, and immersions clusters. Each of these has a different program and a different building design that enhances the interior program. The main idea of the clusters is for people to come together to share knowledge. The programs would be based on the concept of makerspace, in which residents would volunteer to teach different skills or art forms.

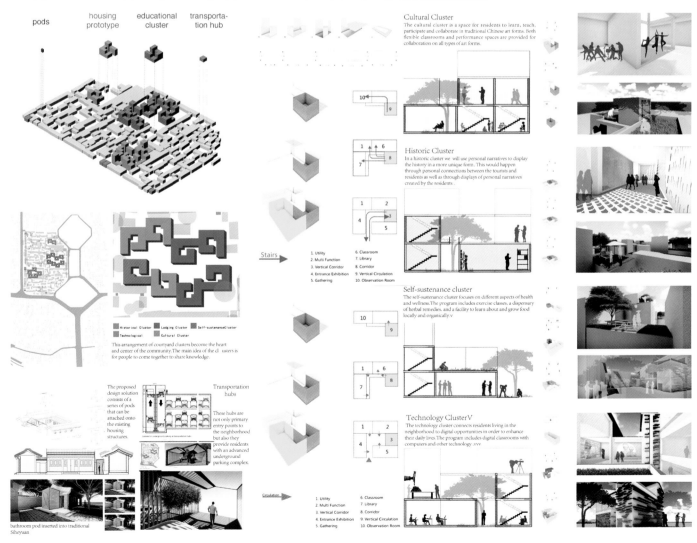

University: North China of Technology
Designer: XueHaoshuo, ChenWanyu, SuYisha, Aikebaier. Uwimana Lydia, Nurlan Babayev, Tinashe Justin Machamire, Emily Williams, Houra Khani, Sergio Gutierre, Juan Galvan, Renzo Pali, Pan Chunguang, Ihumere Irma, Mukazera Marie Christelle
Tutor: Qin Ke, Irma Ramirez, Courtney Knapp, Zhang Bo, An Ping
Course Name: NCUT - Cal Poly College of Environment Joint Design Studio (from grade 4 to grade 5)
Finished Time: Jul., 2015
Exchange Institute: Califonia State Polytechnic University, Pomona

Honorable Mentian

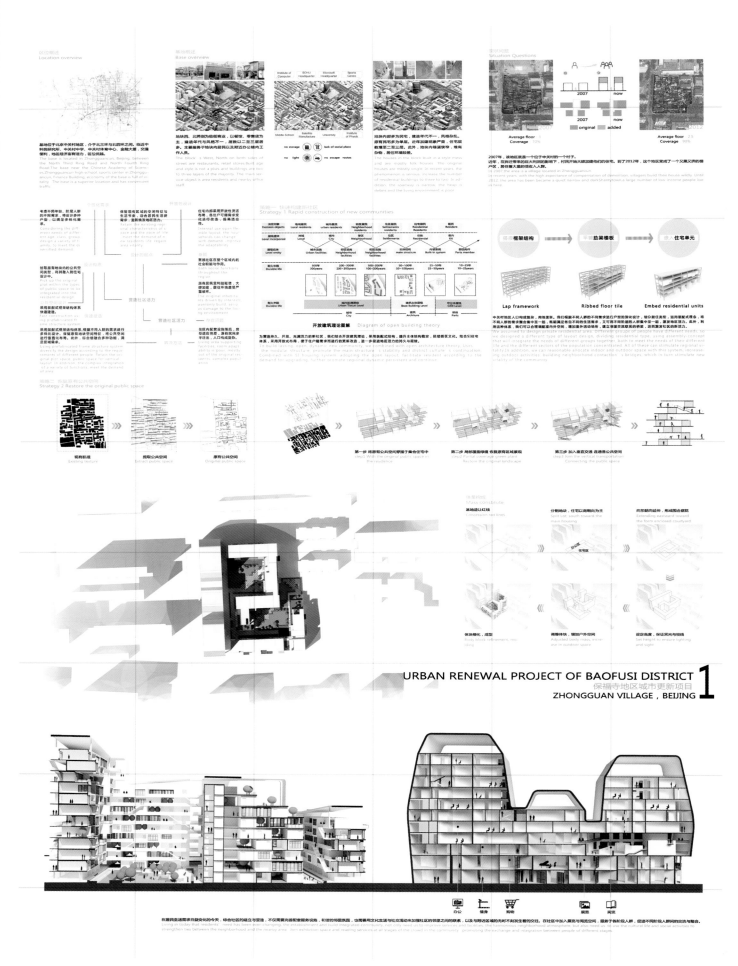

作品名称：保福寺地区城市更新项目（一）
Urban Renewal Project of Baofusi District(1)

入围奖

院校名：北京建筑大学
设计人：王天娇，徐丹，王在书，耿云楠
指导教师：马英，欧阳文，Regine Leibinger
课程名称：柏林工业大学—北京建筑大学—中国建筑设计研究院联合设计课程教学
作业完成日期：2015年05月
对外交流对象：德国柏林工业大学建筑学院

URBAN RENEWAL PROJECT OF BAOFUSI DISTRICT 2
保福寺地区城市更新项目
ZHONGGUAN VILLAGE, BEIJING

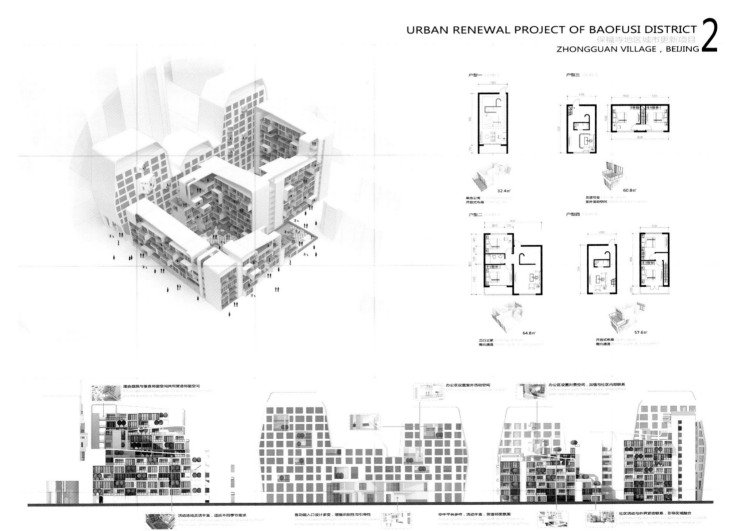

University: Beijing University of Civil Engineering and Architecture
Designer: Wang Tianjiao, Xu Dan, Wang Zaishu, Geng Yunnan
Tutor: Ma Ying, Ouyang Wen, Regine Leibinger
Course Name: Technische Universität Berlin - Beijing University of Civil Engineering and Architecture - China Architecture Design Institute co. Ltd. Joint Design Studio
Finished Time: May, 2015
Exchange Institute: School of Architecture, Technische Universität Berlin

Honorable Mentian

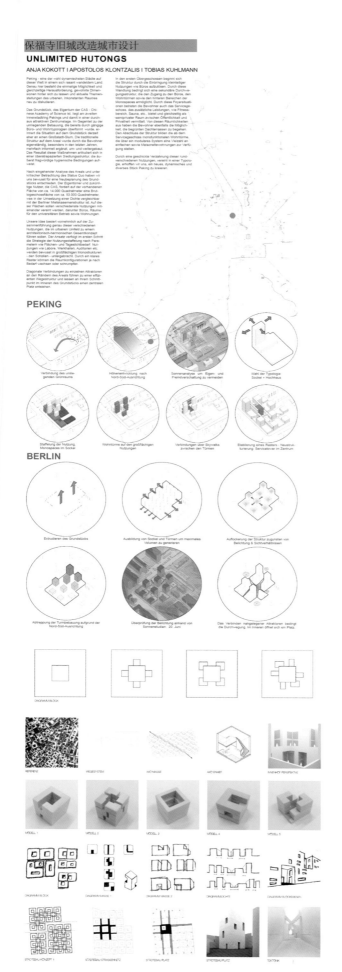

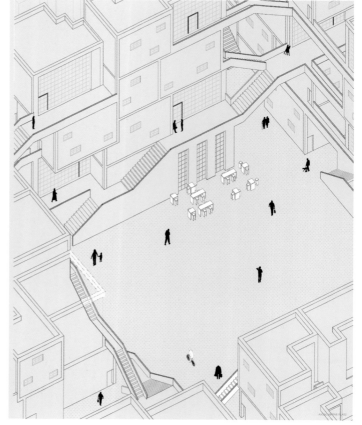

作品名称：保福寺地区城市更新项目（二）
Urban Renawal Project of Baofusi District(2)

入围奖

院校名：北京建筑大学
设计人：杜京伦，魏立志，张明
指导教师：王佐，马英，欧阳文
作业题目：北京建筑大学—柏林工业大学联合设计教学
完成日期：2015年09月
对外交流对象：德国柏林工业大学建筑学院

保福寺旧城改造城市设计
UNLIMITED HUTONGS
ANJA KOKOTT I APOSTOLOS KLONTZALIS I TOBIAS KUHLMANN

We defined ourselves the challenge to increase the density of the given block while at the same time preserving and enhancing the qualities of the grown structure that has been there in the first place. How can you add hight and mass at the same time as maintaining both small scale and programmatic mixture? In that sense we analysed historical building archetypes that offer similar qualities with higher density. As the most appealing reference for our thinking we found the inner city structure of maraqesh to be an advice how dense housing could be organised.

As a result of our analysis we consider a maisonette housing typology with a semi-inner loggia the best interpretation of the traditional hutong housing. These units are grouped blockwise around an inner courtyard. To maintain privacy and high density at the same time these loggias are placed non opposite of each other.

To increase privacy of the housing units we place the circulation out of the blocks. Therefore we introduce an orthogonal grid with crossings to place the circulation staircases and bridges on. By doing that we literally create a built neighborhood to spontanously meet and greet.

The housing blocks connected with their circulation can be densified vertically until their inner ligh- ting gets critical. To maximise the density the most blocks have their lower parts in the south/east or south/west and highest on the opposite. The density also follows an urban skyline, marking a public courtyard wich is created by substracting one block from the middle of the grid. That space is completely surrounded with staircases, allowing for the public to enter office and retail units on multiple levels. These units are switched housing units that are closed to the „housing courtyard" and opened to the „urban courtyard"

In that system the ways of public and private are not predetermined because every staircase or bridge could be the necessary circulation for housing or business, so that is our idea of a positive programmatic mixture.

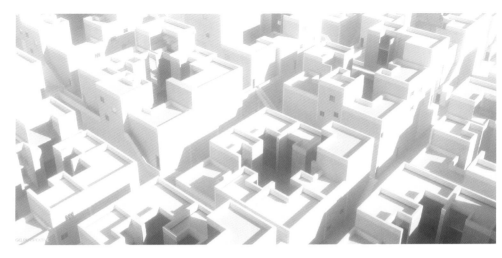

DIAGRAMM EBENEN

CLEAR CUTING REORGANISATION | CONSERVATION AND UND NEW BUILDINGS | CRITICAL URBAN RENEWAL | DIVISION IN BUILDING PHASE | PATHS AND PLACE! | BUILDING PHASE 5

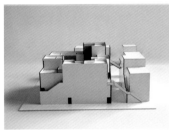

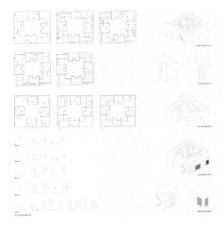 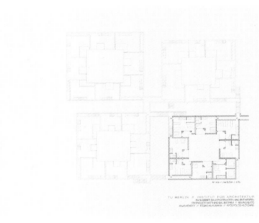

University: Beijing University of Civil Engineering and Architecture
Designer: Du Jinglun, Wei Lizhi, Zhang Ming
Tutor: Wang Zuo, Ma Ying, Ou Yangwen
Coursename: BUCEA-TU BerLin Joint Design Studio
Finishedtime: Sep., 2015
Exchange Institute: School of Architecture Technische Universität Berlin

Honorable Mentian

BACK TO LIFE
Base Position

COMMUNITY CENTER ORIENTED HISTORYIC AREA REGENERATION

Baofu temple area

Bao fu temple village is located in Beijing Chaoyang district is called Chinasilicon valley, zhongguancun area, the surroundin land belongs to the Chinese academy of sciences research off-ice and residential land. Is the 1950 s, Chinese academy of sci-ences in the enclosure movement remaining a rural enclaves.

Today's bao fu temple village, building density, poor health, municipal facilities is not com-plete, no fire control facility and fire escape. Therefore, hai-dian district government is eager to modify the area.

Poor living condition and high security risk——Lack of proper drainage, electricity, firefighting equipment, and lack of enough Evacuation routes

Site Information

current road network | current yard | current building usage | current building height | current building quality

What we do:

Analysis of Map Base | Extract the main roads | Homogenizing treatment | Implanted into traditional form | Enclosing the whole

MASTER DESIGN:
Overall partition

① Community center
② Youth apartment
③ Commercial office
④ Social housing
⑤ Exhibition of leisure
⑥ Community park
⑦ Entrance landscape

REGENERATION CONCEPT:

We from the Chinese traditional garden, four kind of architectural form of building density were studied. As you can see, no matter the east or the west, tradition and modern, the four kinds of different forms of building density has a corresponding match the architectural form.

庭 COURT — Chinese words, is refers to the buildings or surrounded by buildings around the site, that is, a construction of all affiliated sites, vegetation, etc.

院 YARD — Chinese words, house with a wall or fence around the field

园 GARDEN — Refers to the planting flowers and trees for visit places of rest. General garden larger than the school from the scale, and need of artificial landscape.

苑 PARK — Chinese characters, basic meanings: 1. The ancient beast forest planting

Green building analysis:

To simulate the environment of building wind environment and light can be seen from the figure, area building outdoor wind environment is good, the public area air unobstructed, no strong wind area. Great cold, sunshine is meet the requirements, suitable for people in outdoor activities: at the same time improve the building ventilation lighting effect and save energy

wind environment:

	the East	In the modern	THIS DESIGN
庭 COURT			
院 YARD			
园 GARDEN			
苑 PARK			

1.0m

1.5m

3.0m

17.0m

sunshine: Great cold day

作品名称：保福寺地区城市更新项目（三）
Urban Renewal Project of Baofusi District (3)

入围奖

院校名：北京建筑大学
设计人：李取奇，王晓健，牛亚庆
指导教师：晁军，马英
课程名称：保福寺地区城市更新项目
作业完成日期：2015年05月
对外交流对象：德国柏林工业大学建筑学院

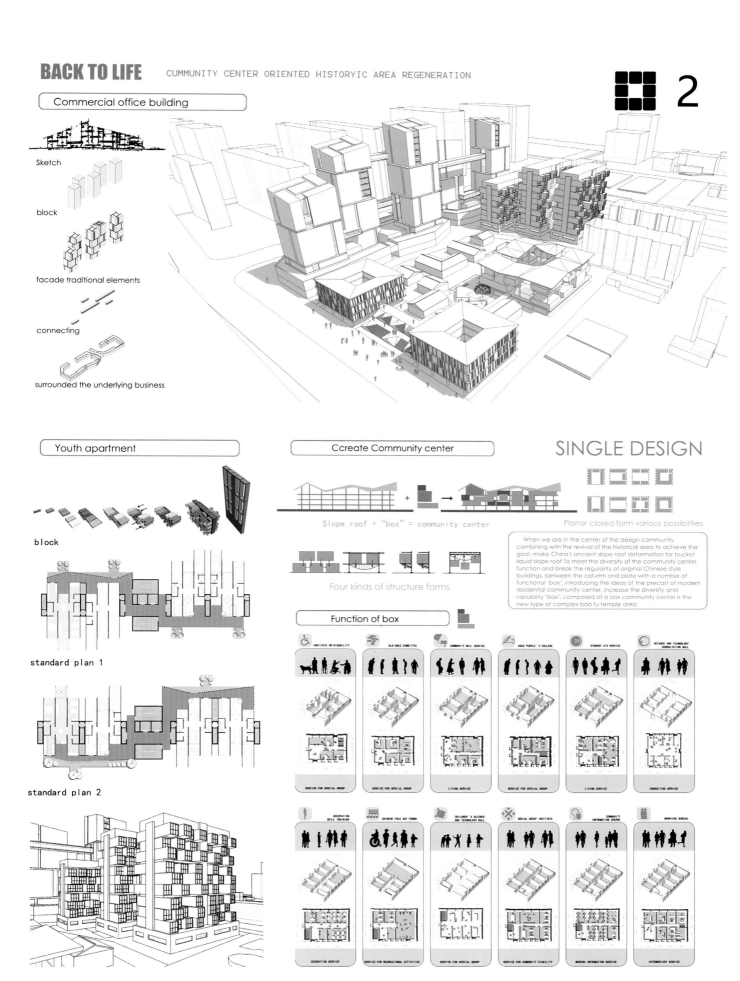

University: Beijing University of Civil Engineering and Technology
Designer: Li Quqi, Wang Xiaojian, Niu Yaqing
Tutor: Chao Jun, Ma ying
Course Name: Urban renewal projects in Baofu temple area
Finished Time: May, 2015
Exchange Institute: School of Architecture, Berlin university of technology

Honorable Mentian

HEARTH CITY HALL

Design Studio
TU Darmstadt
Germany
2015

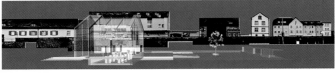

Lindau is an island by Bodensee.

From the island I can see Alps Mountain on the wide water.
So I have an image, that a house built on a wide square.
Glass roof is the shinning snow on the Alps.
Its shape is basic. Like home.
The square is covered by grass, a new node of the park surround the island,
working together as a public space with island hall's square.
The facade is glass, lighting like the famous lighthouse of the island.

I consider that most people walking along the lakeside and autos from the road side.
Its makes the square is made by big stone between two narrow asphalt path.
On the edge of the site, several stone slab raised up.
Working as information boards, and a place to stay a while, between city hall and island hall.
The grass follows the axis of the city hall.
It is like the Japanese temples and the empty base beside them.
Dialog happens between the building and its site.
The stone on the grass is for walking and sitting. The pattern is a shape of Bavaria.

The tree is the very symbol of **Lindau**.

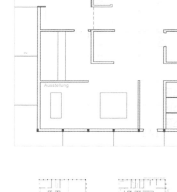
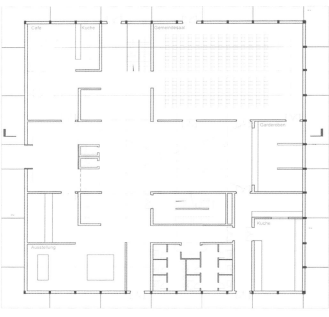

作品名称：博登湖畔市政厅设计
Rathaus am Bodensee

入围奖

院校名：大连理工大学
设计人：尚书
指导教师：Meinrad Morger，王时原，吴亮
课程名称：建筑设计
作业完成日期：2015年07月
对外交流对象：德国达姆施塔特大学建筑系

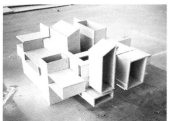 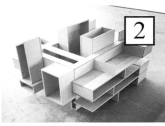

The space inside the building is combined with a light core and three light cubes. Both of them is the place for public service function.

The materials are also basic. Concrete walls, black steel, glass and some coating.
The west facade is a pure glass surface. The frame is steel covered by silicon. Its divides the glass vertically. The horizontal frame is hidden behind the glass.

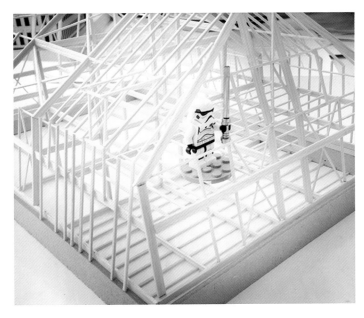
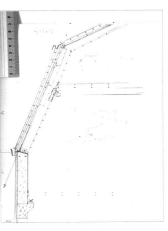 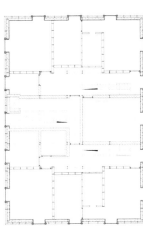

2 Methods of Compositions
The beginning of portfolio comes with 2 composition works in 2011.
The upper one is a composition of entity, but guided by rules of line.
The lower one is a composition started from lines (frame), but presented by entities (shape).

So I realized this is also the point where my works trying to returned. I wonder how architecture happens, how behaviors are translated to places. They are both grammatical and solution.
In fact, the purpose of translation is like literature to language, space exploration to flying. These are all ideas cames from basic function, then ideological pursuits, then higher level functions.

In architecture, I reckon form as entities and ideological pursuit, and technology as lines and functions.

University: Dalian University of Technology
Designer: Shang Shu
Tutor: Meinrad Morger, Wang Shiyuan, Wu Liang
Course name: Surface & ENKO 4
Finished time: Jul., 2015
Exchange institure :TU Darmstadt

Honorable Mentian

都市青年旅館設計

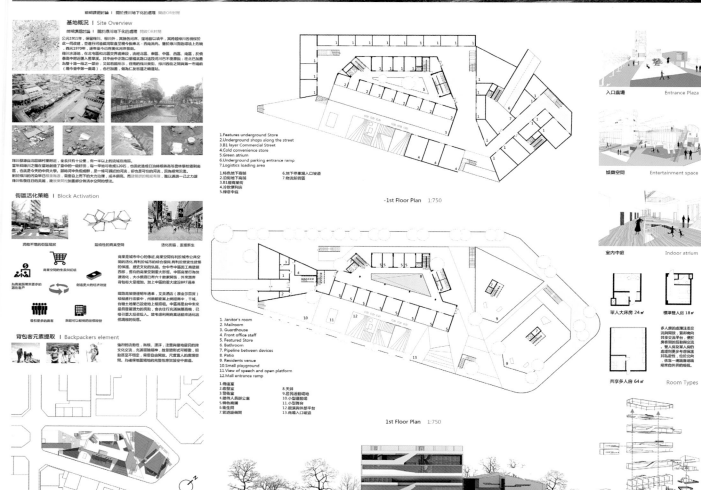

作品名称：青年旅馆设计
Design of Youth Hostel

入围奖

院校名：大连理工大学
设计人：崔培睿
指导教师：杜方中，吴亮，李冰
课程名称：建筑设计（七）
作业完成日期：2014年12月
对外交流对象：逢甲大学

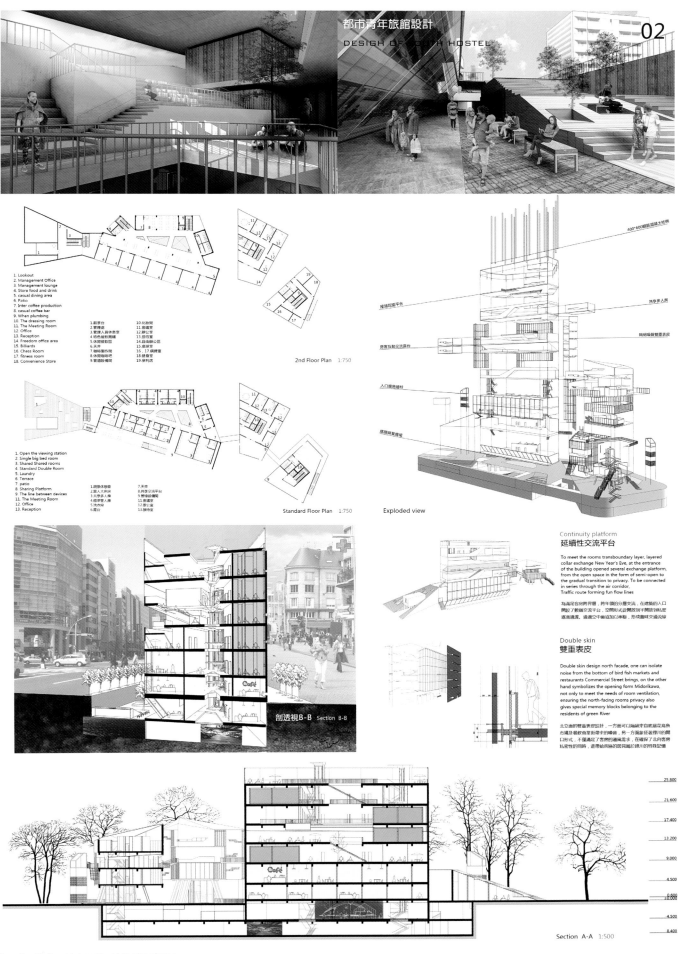

University: Dalian University of Technology
Designer: Cui Peirui
Tutor: Du Fangzhong, Wu Liang, Li Bing
Course Name: Design Studio VII
Finished Time: Dec., 2014
Exchange Institure : Feng Chia University

Honorable Mentian

SHUTTLING
PYEONGHWA MARKET REUSE

01

SITE PLAN

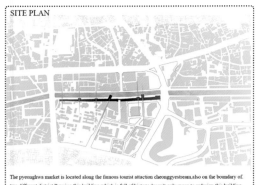

The pyeonghwa market is located along the famous tourist attaction cheonggyestream,also on the boundary of two different district.Reusing this building which is full of history doesn't only mean to redesign this building along,but also to combine the old building with new social life ,new requirements of city ,and the constantly changing modern life.

The architecture is divided into two general parts: one is the rational and efficient area which is composed of different kinds of shopping staffs,exhibition area and design studios ,the other is the intentionally enlarged transportation space,which is an atrium with ramp ways connecting with different storeys.

While people shopping ,in this building they shuttle between the old market and the new atrium,the history and new identity of the city,the memory and the present life of fashion seoul.

HISTORY LINE

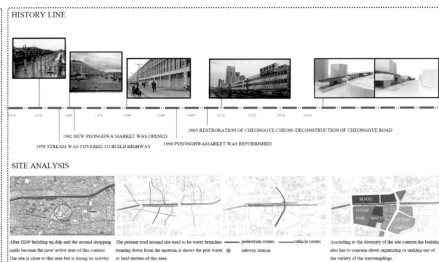

1962 NEW PEONGHWA MARKET WAS OPENED   2005 RESTRORATION OF CHEONGGYE CHEON/ DECONSTRUCTION OF CHEONGGYE ROAD
1958 STREAM WAS COVERED TO BUILD HIGHWAY   1998 PYEONGHWAMARKET WAS REFURBISHED

SITE ANALYSIS

After DDP building up,ddp and the around shopping malls become the most active area of this context. The site is close to this area but is losing its activity little by little.

The present road around site used to be water branches running down from the mountain,it shows the past water or land texture of this area. pedestrian routes vehicle routes subway station

According to the diversity of the site context,the building also has to concern about organizing or making use of the variety of the surroundings.

PROGRAM

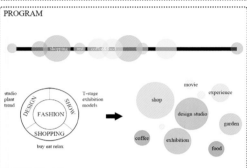

FASHION PLANT

Here you can find anything about fashion!
Pyeonghwa market used to be a wholesale market which shops and plants are combined together.According to the context,the site is surrounded by popular shopping places in seoul.Also,in the north of the building ,there are stream and the primitive east gate of seoul city..Then the new program is supposed to be a complex of fashion shopping and sightseeing.

Whole building is divided into two parts,and the atrium of ramps is a good place for sightseeing .As for the fashion shopping program,it consists everything of fashion,from the source of fashion----design to fashion publicity -----exhibition and finally get the own fashion of yourself----shopping.
Other programs such as restaurant ,coffee shop,punlic garden,movie theatre are serving fashion

150-200m is the distance people will get bored and tired.In order to avoid it happening ,some active nodes are inserted in the whole shopping routes.

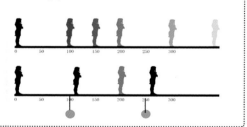

CONCEPT DIAGRAM

three-storey,six hundred-long building   divide the building into two parts north part is transparent as atrium for sightseeing south part is solid for real shopping function   in north part ramps are used as circulation among different storeys in south part the original structure is reserved as an efficent shopping space   insert interesting "nodes"---functional boxes into the south part and north part

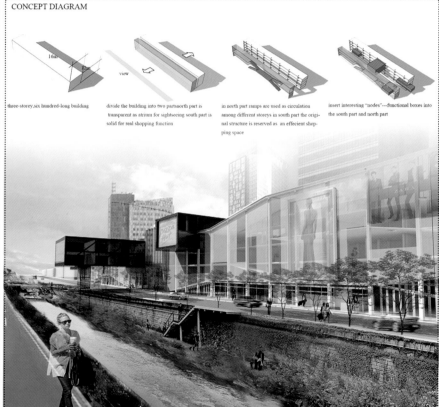

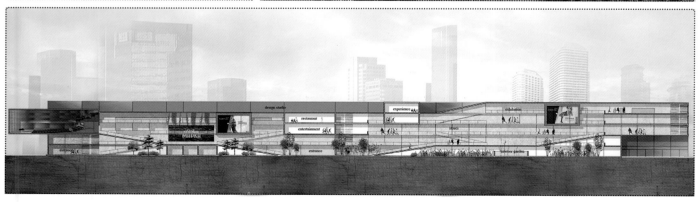

作品名称：和平市场改造计划
Pyeonghwa Market Regeneration

院校名：大连理工大学
设计人：徐佳臻
指导教师：Fabio Dacarro，李冰，吴亮
课程名称：建筑设计 Ⅴ
作业完成日期：2014 年 12 月
对外交流对象：韩国高丽大学建筑系

入围奖

SHUTTLING
PYEONGHWA MARKET REUSE

02

DETAIL OF BOXES

Because this is an old building reuse design, how to deal with th old structure is the first important probloem we should study.
According to many times site vist, we found that Pyeongyang market is frame structural system, and the most of the columns are in bad condition, but northside columns are in very bad condition, so we decide to use these strategy to solve the structural problem.
— break down the norhtside columns and replaced with new steel columns
— use steel to reinforce old columns and ceilings
— choose truss structural system to build the insert boxes

- original columns
- reinforce steels
- new truss structure
- reinforcement

SECTIONS

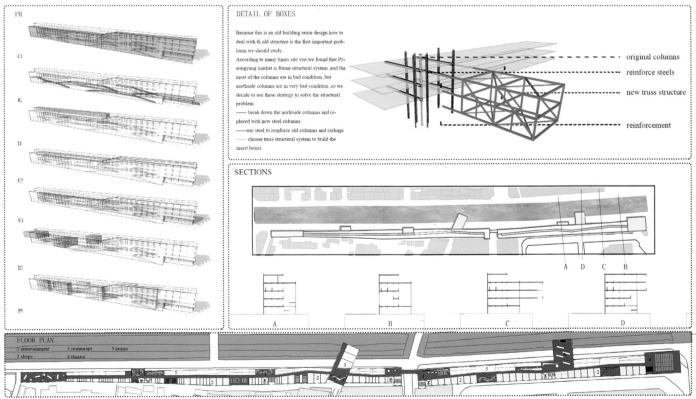

University: Dalian University of Technology
Designer: Xu Jiazhen
Tutor: Fabio Dacarro, Li Bing. Wu Liang
Course Name : Design Studio V
Finished Time: Dec., 2014
Exchange Institute: School of Architecture, Korea University

Honorable Mentian

社区建筑　学习科学书院设计

本课题是东南大学以及东京工业大学住宅设计联合教学。本教学旨在通过研究建筑类型学，理解空间特征，并通过学习空间结构的设计方法，处理在复杂的中国城市环境下有一定特征的高密度社区建筑设计。基地位于东南大学四牌楼校区南门沙塘园宿舍地块，需要处理与东南大学的老建筑以及周围社区之间的关系。最终设计出，具有开放性并带有城市特征的社区建筑。

本设计通过研究学习科学中心书院空间特征，首先设计出供儿童使用的开放平台以及供研究人员观察使用的大空间，从而产生十字形的空间结构，用以应对沙塘园和老房子之间的空地。紧接着通过设计其余功能的体量，从而让十字形结构影响到整个建筑群。最终设计出具有开放性并带有城市特征的高密度社区建筑。

Site Area:7900m²
Buiding Area:13288m²
Coverage:46%
FAR:1.68
Parking Capacity:43

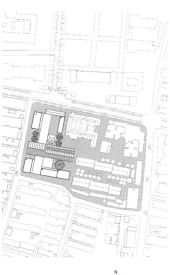
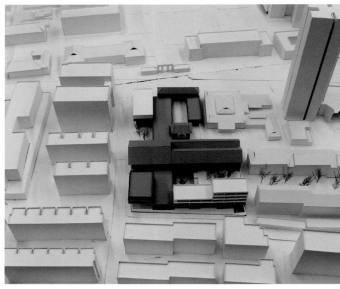

Site 1:2000

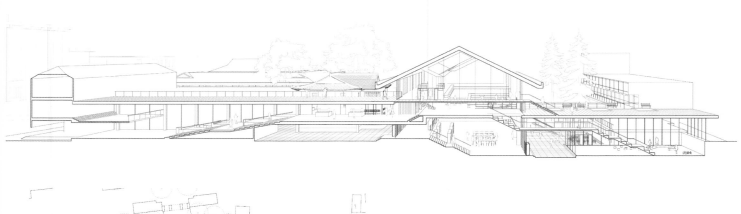

Ground plan 1:600

First Floor plan 1:600

Underground plan 1:600

作品名称：社区建筑——东南大学书院设计
Community Building

院校名：东南大学
设计人：沈忱，隋明明
指导教师：葛明，奥山信一
课程名称：东南大学—东京工业大学联合设计课程教学
作业完成日期：2015年08月
对外交流对象：日本东京工业大学建筑学院

入围奖

COMMUNITY BUILDING

This joint course is cooperated by Southeast University and Tokyo Institute of Technology. In order to design a high density community building located in a complex Chinese urban surroundings, we studied the design method of space structure and space characteristics, by learning the architecture Typology. The site is located in the community at the opposite of the south entrance of Southeast University. The course need students deal with the relationship among the building to be designed, other existing buildings and surrounding environment.

By studying the characteristics of RCLS academy, firstly, the students designed an exoteric platform for the living in children and a space for researchers, and with the help of which, a cross-shape space structure was designed to deal with the relationship with Shatangyuan and the inner garden between two old buildings. Then, after other parts of the building being designed, the cross-shape space structure can affect all the parts of building.

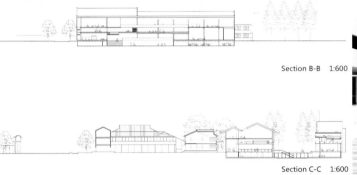

Section B-B 1:600

Section C-C 1:600

University: Southeast University
Designer: Shen Chen, Sui Mingming
Tutor: Ge Ming, Shinichi OKUYAMA
Course Name : Southeast University - Tokyo institute of technology Joint Design Studio
Finished Time: Aug., 2015
Exchange Institute: School of Architecture, Tokyo institute of technology

Honorable Mentian

适应性住区设计

设计要求学生发展一个随时间和空间变化的满足个性化需求的建筑概念。

建筑的问题是满足使用者的需要。每个使用者的需求不一样，即使是同一个使用者，需求也会随时间变化。如果建筑的目的是服务于使用者，建筑就要在不拆除毁坏建筑，满足可持续的条件下，适应这些变化。

这个项目涉及2个建筑问题：

时间：确保建筑的可持续性，在建筑服务期限内，和在结构允许的限度内，灵活适应使用功能和环境等不可预测变化。

人：现实的人的需求不是统计数据，功能或设计计划能替代的。如果多元化这个词是今天人的行为特征，那么，没有建筑与人的互动，就不可能满足人的需求。

本设计课程的引入积极的，真实的，动态的和活力十足的空间设计手段和方法。这一方法满足日常生活行为，气候和需求的改变。持续不变的事件和结构之间的互动将导致形态和空间的积极转化。建筑并不是纪念碑，而是操作和运营的过程。设计的目标并不是解决最终的问题，而是一个包括从场地条件分析研究到转化过程。这些条件包括：开放性结构，复杂性，中性，整合性，临时性，移动性，基础结构性的而不是建筑性的等等。

1. 一个生活单元的可移动系统
a movable system for a living unit

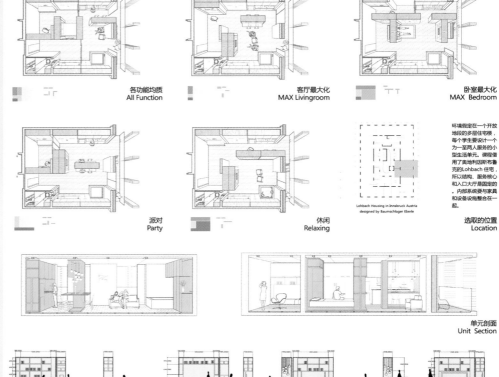

各功能均质 All Function
客厅最大化 MAX Livingroom
卧室最大化 MAX Bedroom
派对 Party
休闲 Relaxing
选取的位置 Location

环境假定在一个开放地段的多层住宅楼，每个学生要设计一个为一至两人服务的小型生活单元。课程借用了奥地利因斯布鲁克的Lohbach住宅，所以结构，服务核心和入口大厅是固定的。内部系统要与家具和设备设施整合在一起。

单元剖面 Unit Section

灵活构件模数化 Modulization of Elements

2. 只需有外立面，交通，结构的住宅
a building with façade, circulation, and structure only

场地的总平面设计由工作小组共同协商完成，试图在讨论建筑体量的同时，关注外部空间的品质和魅力，为不同的市民活动提供可能性，以满足不断变化的使用要求。

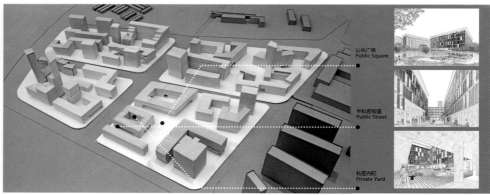

公共广场 Public Square
半私密街道 Public Street
私密内院 Private Yard

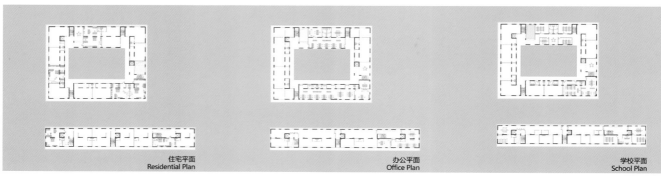

住宅平面 Residential Plan
办公平面 Office Plan
学校平面 School Plan

作品名称：住区设计
Adaptive Residence

入围奖

院校名：东南大学
设计人：李鸿渐
指导教师：贾倍思
课程名称：建筑设计IV
作业完成日期：2015年01月
对外交流对象：香港大学建筑学院

ADAPTIVE RESIDENCE

The students are invited to develop an architectural concept addressing individualization and change through space and time.

Any architectural settlement should basically meet the user needs. Those needs are different from one user to the other. And the needs of each user change through time. If the building wants to serve its occupants, the building has to accommodate change as well, without any destruction or demolition according to the sustainability agenda.

The studio program invites ideas of the active, lively, dynamic, energizing spatial instrument which is transformative responding to the change of activity, climate, need or purpose in every day life. Architecture is not monument, but the process of operation.

The studio calls for proposals of an instrument in constant movement, strategically, and technically offering form and space to be undetermined, undersigned, open for mutation in time, and open for interpretation, again with physical interaction with people.

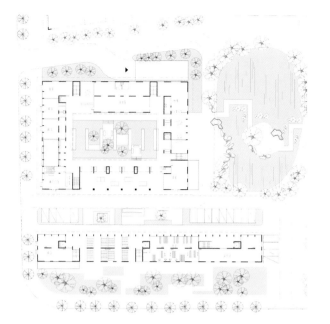

顶层平面
Top Floor Plan

一层平面
First Floor Plan

地下层平面
Underground Plan

3. 可移动体系的置入及空间深化
introduction of movable system and space development

A- 户型平面
Unit A Plan

B- 户型平面
Unit B Plan

C- 户型平面
Unit C Plan

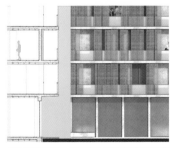
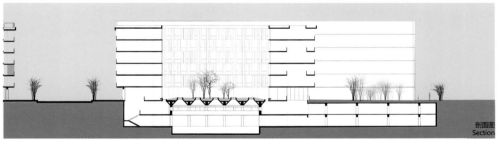

剖面图
Section

University: Southeast University
Designer: Li Hongjian
Tutor: Jia Beisi
Course Name: Adaptive Residence Design
Finished Time: Jan., 2015
Exchange Institute: School of Architecture, Hongkong University

Honorable Mentian

波士顿音乐广场

本课题为中美联合毕业设计。从北面的滨水海岸到南面的城市环境，我们选择了用两边公共开放，中间专业封闭的办法来过渡。为了衔接在四号码头处断裂的滨水漫步道，我们留出了充裕的漫步空间和多样的选择供步行者选择。同时还设立了水上出租车的停靠站点，直接接待来自机场的游客和货物。BMP和ICA互补的主题策划让两所文化建筑共同形成了创意新区的文化中心。这两座建筑合抱着一片繁忙的水域并且共享了东面的一片服务区域。

作品名称：波士顿新区音乐广场设计
Boston Music Playground

院校名：东南大学
设计人：马斯文
指导教师：鲍莉
课程名称：毕业设计
作业完成日期：2014年12月
对外交流对象：美国艾奥瓦州立大学

入围奖

BMP (Boston Music Playground)

From the waterfront south coast to the north of the city environment, we chose to use public open on both sides, intermediate professional closed way to transition.

To cohesive fracture at Pier 4 waterfront walk, we walk out the abundant space and a variety of choice for the pacers to choose from. Also set up a water taxi stops at the same time, the reception of visitors from the airport and cargo directly.

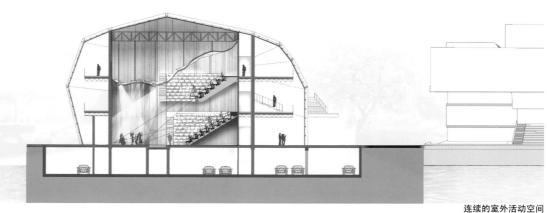

连续的室外活动空间

亲水层
地面层
屋面层

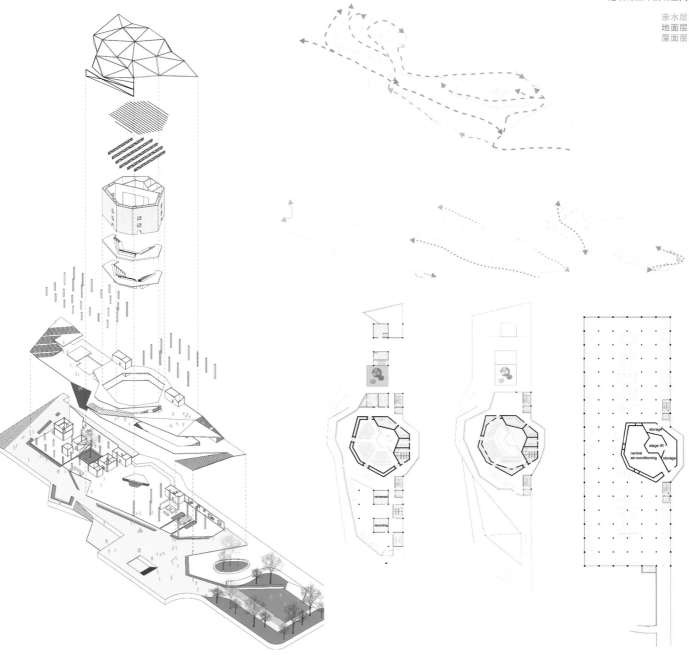

University: Southeast University
Designer: Ma Siwen
Tutor: Bao Li
Course Name: Iowa State - Southeast University Joint Design Studio
Finished Time: Dec., 2014
Exchange Institute: College of Design, Iowa State University

Honorable Mentian

Rooftop Hiking

Inter-generational Community — General Planning & Architecture Design (Senior living health & wellness, rehab, clinic)

Master Plan 01

Background

Aging has become a global issue. Both the United States and China have rapidly increasing elderly population. By 2020, both countries will have more than 20% of elderly population. How to design the physical environment for elderly to stay involved in the society, and engage is physical and social activities is a challenge to both countries. In the U.S., different approached have been explored, including independent living, assisted living, nursing home, and CCRC (Continuing Care Retirement Community). However, the above design solutions are based on age-restricted communities. Older people are isolated from the other age groups. Recently, there is growing trend of developing inter-generational community that is supportive of cohousing of various age groups. The success of an inter-generational community is highly dependent on the health care and service support provided through the community. It is imperative to provide activity, interest, excitement, and connections to support a dynamic community in which people age well.

The site is the lot next to the Rock Chalk Park, a recently opened sports and recreational center. This size of the lot is about 142.33 acres. The intention is to take advantage of existing sport facility in Rock Chalk Park, and create a community that knits together the functional programs of therapy and fitness so residents are motivated and inspired to engage in physical activities. The other intention is to help activate the area around the Rock Chalk Park through this project. The area is expected to act as a bridge to existing neighborhoods and help enhance physical activities and social interaction, therefore to develop a sense of community.

Growth of City

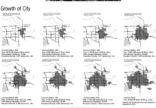

Reference

City Adjacency

Traffic

Facilities

Site Photos

Topography

Section 1 — Landscape / Health Care / Independent Living
Section 2 — Landscape / Culinary / Education
Section 3 — Landscape / Independent Living / Landscape / Independent Living / Sports

Master Plan

Concept

This area is a little far away from the downtown of Lawrence, and there are not many facilities which are designed for the elder around it. Besides, all the bus stations are a little far away for the elder because they go out mainly by walking. It is really not suitable for the elder to live a convenient life here.

To change such situation, we take the block size of the downtown as reference, and build a main street for pedestrians to have entertainments.

We also set different functions into different blocks so that these blocks can appeal to different types of the elder.

About the health and wellness center, we choose two blocks located on the east side of this area as the site of it bacause of the following reasons:

1. it is near the center main street, and can be reached easily, so that it can cover all the area;

2. it is also not far away from the main entrance, so that it can also provide health service for the residence around.

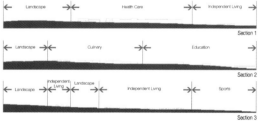

作品名称：年龄混合型的老年居住、健康、医疗、康复社区规划与建筑设计

Inter-generational Community : Serior living, Health and Wellness, Rehab, Clinic

入围奖

院校名：南京工业大学
设计人：梁末，季鑫
指导教师：蔡志昶，方遥，Hui Cai, Kent Spreckelmeyer
课程名称：联合毕业设计
作业完成日期：2015 年 01 月
对外交流对象：美国堪萨斯大学建筑学院

Rooftop Hiking

Inter-generational Community   General Planning & Architecture Design (Senior living health & wellness, rehab, clinic)   Architecture Design 02

Site Plan

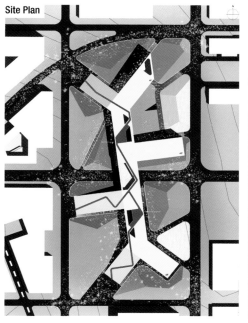

How to deal with two sites?

Separate
Dealing with the two sites separately can make it easier to manage the flow of the elderly and the illness. However, it is inconvenient for the elderly to get the treatment in time if they need help.

Linkage
Since the requirements of the wellness center and the hospital are similar in some way. Both the elderly and the illness need health care. Taking the similarity into consideration, I decide to use the building to link the two sites.

Process Diagram

Connection
Since the topograghy is a little cliffy in the site, I choose the strip block to connect the two parts of site.

Rotation
For the aim that reduce the damage to the site, I rotate the block, so that the orientation of the building can follow the topography.

Stretch
In consideration of the different kinds of entrance and the environment around the site, I stretch the block to different orientation.

Distortion
The west part of the site is higher than the east, so I pull the west fingers of the building down, which will make the roof accessable.

Access
The main methods for the elderly going out is walk. I separate the routes of human and vehicle. The main roadsconcentrate on the east of the site, and the paths forthe elderly are located on other three sides.

Landscape
The landscape forms pockets of different character between each finger of the building. The pockets become attractive out-door spaces, courtyards and recreation areas.

Rooftop Hiking Path
I propose a public trail leading over the peak of the community health center. The peak will become the destination for locals and visitors and provide panoramic views of the landscape.

Pedestrian Circulation   Hiking Path   0.000m Plane   4.500m Plane   9.000m Plane   13.500m Plane

South Elevation   North Elevation

East Elevation

University: Nanjing Tech University
Designer: Liang Mo, Ji Xin
Tutor: Cai Zhichang, Fang Yao, Cai Hui, Kent Spreckelmeyer
Course Name: Nanjing Tech University - Kansas University Joint Design Studio
Finished Time: Jan., 2015
Exchange Institute: Kansas University School of Architecture

Honorable Mentian

City Wall Lane Area Transformation and Connection Design I

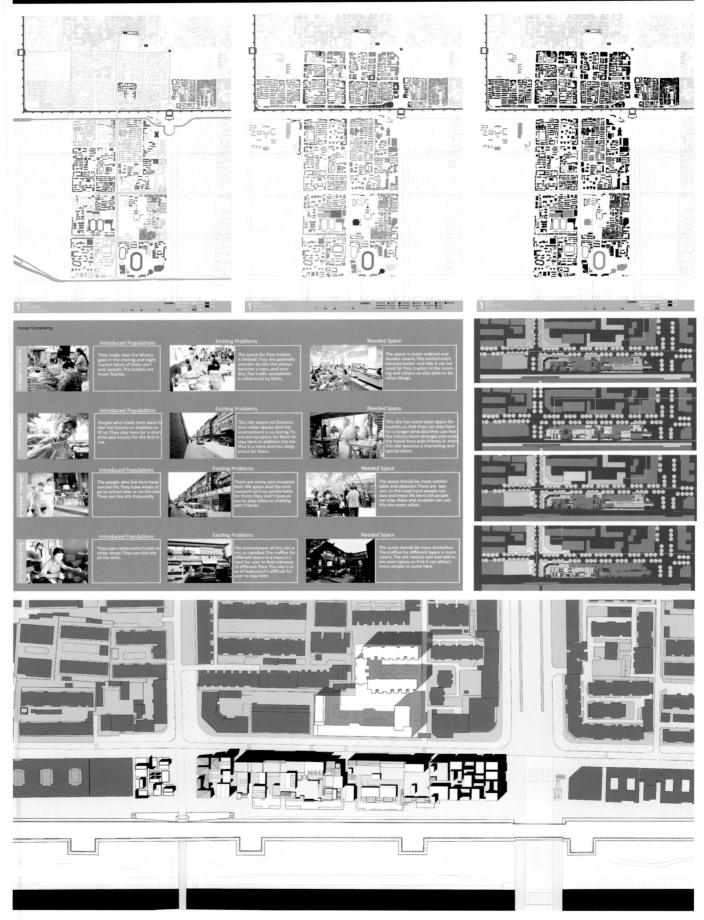

作品名称：转变·演绎——唐轴线朱雀门顺城巷地段城市设计
City Wall Lane Area Transformation and Connection Design

入围奖

院校名：西安建筑科技大学
设计人：方坚，徐娉，Alessandra Guizzi, Giulia Dagheti, Greta Bosio, 武琼
指导教师：李昊，常海青，鲁旭，李煜，Laura Anna Pezzetti, Carlo Palazzolov
课程名称：2015 西安建筑科技大学—意大利米兰理工大学国际联合工作营课程教学
作业完成日期：2015 年 05 月
对外交流对象：意大利米兰理工大学建筑学院

City Wall Lane Area Transformation and Connection Design II

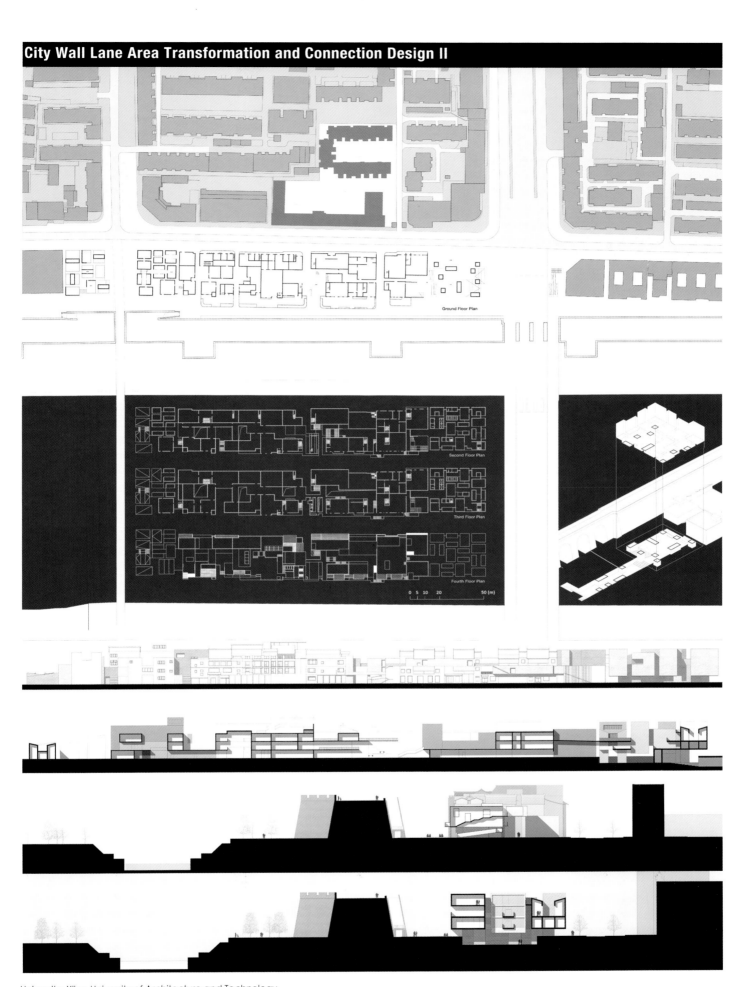

University: Xi'an University of Architecture and Technology
Designer: Fang Jian, Xu Ping, Wu Qiong, Alexader, Greta Bosio, Guilia Gagheti
Tutor: Li Hao, Chang Haiqing, Lu Xu, Li Kun, Laura Anna Pezzetti, Carlo Palozzolo
Course Name: 2015 Polimi - XAUAT Interenational April Workshop
Finished Time: May, 2015
Exchange Institute: College of Architecture, Politecnico di Milano

Honorable Mentian

CONTINUATION BY WALK 行走--延续

2015 " CHINA & AMERICAN JOINT TEACHING " COURSE DESIGN RESULTS

A MUSEUM DESIGN PAGE 1

The main spatial driver of our project comes from the axis of the ancient Wei River Bridge as a metaphor of transportation between modern and traditional forms of expression, as well as the literal axis between the city and the surrounding area. Based off of the Chinese character of "center," our group diagrammed the project by setting each word group spatially opposing around the character, with the bridge representing the axial strike through the middle of the "center."

CONCEPT
概念

In addition to the spatial association made with the "center" character, the proposal also connects to the multiple temporal stages of the site to compose the scheme that forms our project. We are trying to respect and represent the spatial arrangement of the bridge as it was in the past; draw a connection to the bridge as the spatial crux between the tombs of the Han Dynasty and the Han city; respond to the modern surroundings and the forecasted population growth in the area; and celebrate the current archaeological status that the site will partially continue. To accomplish all of this, we are proposing a dynamically interactive exhibition of the bridge piles and ship wreckage that functions as a public space and educational facility to spatially organize the surroundings now and for the future.

HISTORY BACKGROUND
历史背景

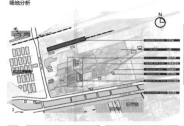

SILK ROAD FIRST BRIDGE
丝绸之路第一桥

BRIDGE CONSTRUCTION
FLAGSTONE
SOIL
PLANK
PILLAR
RIVERBED

BRIDGE MEMBER
BYPASS FLOW

SITE ANALYSIS
场地分析

SITE BOUNDARY
Since the base is located in the city to grow and spread in the region, the base should be designed to give it more life functions in the futher.

RAILROAD STATION
Railroad station is an important element within the base, should be properly deal with its transport links and environmental immunity.

SURROUNDING ROADS
On the south side of the base is the city expressway,as to avoid interference between the base, it should be set up some buffer space.

ANCIENT PILLAR RUINS
The main point of the ancient ruins of the Chucheng gate bridge, need the necessary protection mesures.

ANCIENT BRIDGE RUINS
It should follow the axis formed by the Chucheng gate bridge, response to the history, shape the design of the bridge of the clever use of the form.

HILLSIDE
Utilization of the hillside, which was transformed into green space, and create a base for a green environment.

SURROUNDING DESIGN
场地环境设计

1. Freight Station
2. Bridge of Unreal
3. Floral Park
4. Station Square
5. Cycling Park
6. Expo Center
7. Cultural Park
8. Experience Crop
9. Events Plaza
10. Entrance Plaza
11. Parking

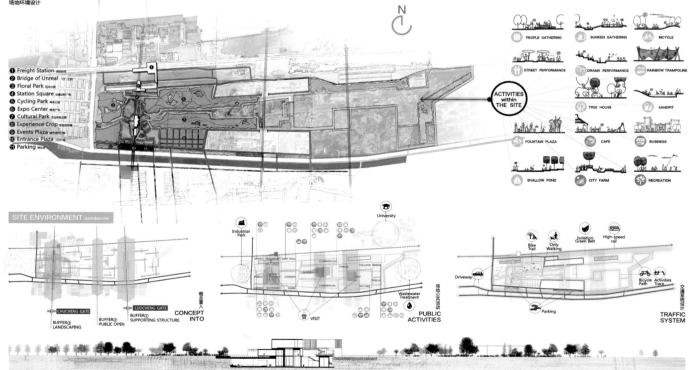

CROSS SECTION

作品名称：行走—延续——西安渭桥遗址区博物馆设计
Continuation by Walk

入围奖

院校名：西安建筑科技大学
设计人：陈哲怡，崔筱曼，Wendy Stradley, Kinga Pabjan
指导教师：常海青，苏静，同庆楠，吴涵儒，鲁旭，Albertus Wang, Martin Gold
课程名称：2015中美联合设计课程教学
作业完成日期：2015年01月
对外交流对象：美国佛罗里达大学建筑学院

CONTINUATION BY WALK 行走--延缩

PAGE 2

The Axis exists so that the casual observer can stand within our museum, and understand it from one perspective-only to look back and see a derivation. The observer then wanders through continuity of the skewed path until he reaches the other side. Only then can the visitor have a complete knowledge of the bridge remains as a spatial organizer between the historical and the current.

TOTAL GRAPHIC DESIGN
基地总平面设计

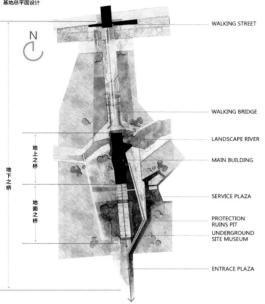

- WALKING STREET
- WALKING BRIDGE
- LANDSCAPE RIVER
- MAIN BUILDING
- SERVICE PLAZA
- PROTECTION RUINS PIT
- UNDERGROUND SITE MUSEUM
- ENTRACE PLAZA

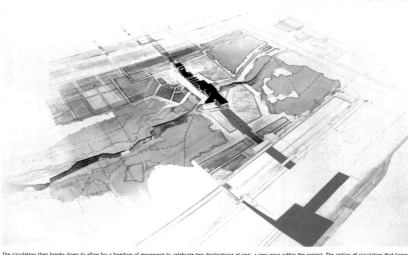

The circulation then breaks down to allow for a freedom of movement to celebrate two destinations at one's own pace within the project. The option of circulation that lowers visitors into the archaeological pit breaks immediately off of the entryway. This path then leads the observer around immediately to the first glance of the pillars from the side, then to the ship wreckage unique to this site in Xi'an. The visitor then can go outside into the archaeological pit to see the rest of the bridge piles, which would be encapsulated closely in glass to encourage an interactive meandering through the outdoor area. Following a loosely-defined path around the pillars, the visitor would then go through a showcase of the objects found in the Wei River excavation sites, and then back up into the main body of the structure. The next option of circulation, which can be chosen either before or after the previously mentioned prescribed path, would circulate first through the main educational exhibits of the exam, showing the relationship between the bridge, the Han city, and the modern city of Xi'an.

ANALYTICAL PROGRAMS
方案解析

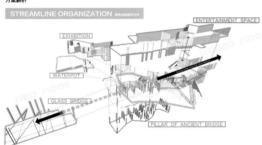

BUILDING PLAN
建筑平面

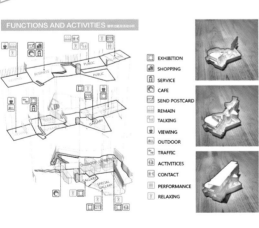

- EXHIBITION
- SHOPPING
- SERVICE
- CAFE
- SEND POSTCARD
- REMAIN
- TALKING
- VIEWING
- OUTDOOR
- TRAFFIC
- ACTIVITICES
- CONTACT
- PERFORMANCE
- RELAXING

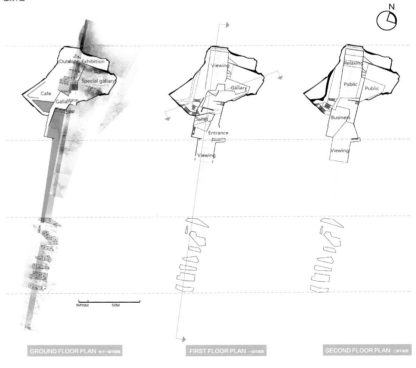

GROUND FLOOR PLAN | FIRST FLOOR PLAN | SECOND FLOOR PLAN

LONGTITUDINAL SECTION

University: Xi'an University of Achitacture and Technology
Designer: Chen Zheyi, Cui Xiaoman, Wendy Stradley, Kinga Pabjan
Tutor: Chang Haiqing, Su Jing, Tong Qingnan, Wu Hanru, Lu Xu, Albertus Wang, Martin Gold
Course Name: Xi'an University of Achitacture and Technology - University of Florida Joint Design Studio
Finished Time: Jan., 2015
Exchange Institute: University of Florida

Honorable Mentian

2015 | SCUT + U.C.Berkeley

THE BREWERY: ADAPTING FOR A HEALTHIER FUTURE
URBAN DESIGN WORKSHOP OF GUANGZHOU PAZHOU WATERFRONT
广州琶洲滨水区城市设计工作坊

When we come to the issue about the post-industrial development of the site, Pearl River Brewery, we start with studies on the conditions of the warehouses. And according to the studies, warehouses in the middle and the north are better to remain and reuse. On the other hand, in order to help the site develop sustainably, we introduce different themes for different part of the site, including LIVE, INNOVATE, CREATE and ABSORB / GROW. Each part could create value for the site and INNOVATIVE and CREATIVE parts are reusing the warehouses in the brewery. By applying various methods on architecture and landscape design, we are trying to make it a sustainable development in terms of human happiness, ecology and economic.

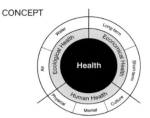

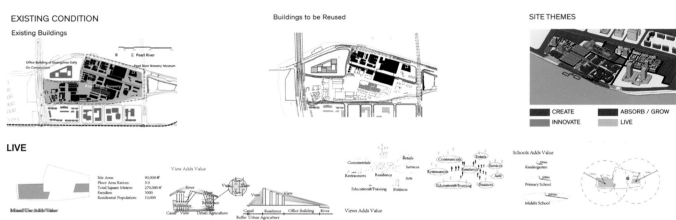

LIVE
By taking down the warehouses in bad condition in the south and in the east, we purpose to build new residential buildings there to creat value economically for the site.

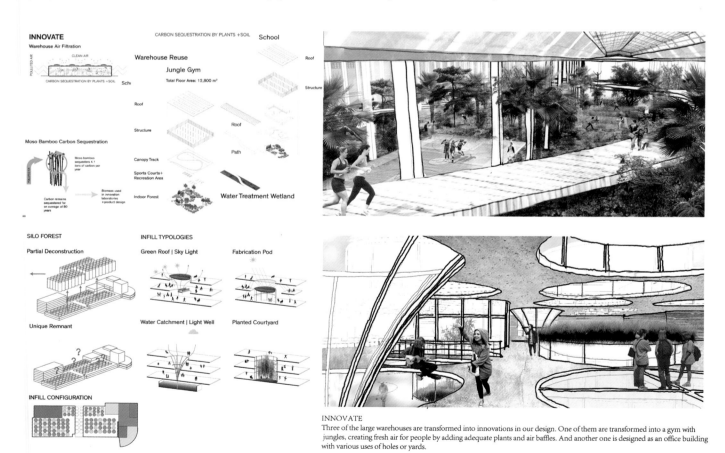

INNOVATE
Three of the large warehouses are transformed into innovations in our design. One of them are transformed into a gym with jungles, creating fresh air for people by adding adequate plants and air baffles. And another one is designed as an office building with various uses of holes or yards.

作品名称：啤酒厂改造
Reconstruction of Brewery

院校名：华南理工大学
设计人：梁雅晴，李越宜，司竞一，吴百荣
指导教师：孙一民，Peter Bosslmann，苏平，王璐，周毅刚，李敏稚
课程名称：SCUT-UCB 联合工作坊琶洲城市设计
作业完成日期：2015 年 01 月
对外交流对象：美国加利福尼亚大学伯克利分校建筑学院

入围奖

2015 | SCUT + U.C.Berkeley

THE BREWERY: ADAPTING FOR A HEALTHIER FUTURE
URBAN DESIGN WORKSHOP OF GUANGZHOU PAZHOU WATERFRONT
广州琶洲滨水区城市设计工作坊

CREATE

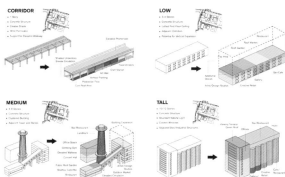

CREATE
By transforming the existing factory buildings in several ways, including reinforcing, expanding and building new constructions, we aimed at changing the create zone into a lively public centre with amusement and creative function attached.

ABSORB / GROW

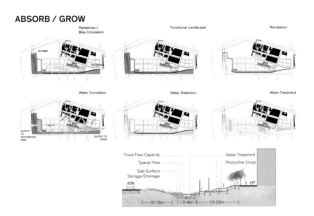

ABSORB
Based on a existing river, we hope that the absorb zone could be a buffering centre with local agriculture combined urban landscape, for absorb and adjust the influences from the surrounding zones such as the 40-meter highway, tall office buildings and create and amusement centre.

PLAY

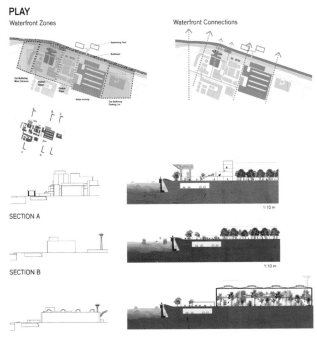

PLAY
With a ferry dock inside this zone and a LRT train line acrossed, we proposed this play zone to be a integrated active zone allowing regular routine commute activities happen as well as the occasional healthy social ones.

University: South China University of Technology
Designer: Liang Yaqing, Si Jingyi, Wu Bairong
Tutor: Sun Yimin, Peter Bosselmann, Su Ping, Wang Lu, Zhou Yigang, Li Minzhi
Course Name: Urban Design Workshop of Guangzhou Pazhou waterfrom
Finished Time: Jan., 2015
Exchange Institute: U.C.Berkeley

Honorable Mentian

T.I.T Urban Design Strategies Reserch

01

T.I.T, GUANGZHOU, CHINA

Looking at TIT Creative Industry Zone means looking at a part of the history of Guangzhou and then, of China. The chapter is structured as sequence of jumps in time, space and scale. Thinking about a complex intervention on industrial heritage within on-going urban development processes requires a new set of strategies that could be addressed only by rethinking our own design tools. Working on different scales, comparing cases in different contexts, contrasting synchronic and diachronic views from the Small to the Extra Large dimension is our way to broaden our understanding and refine our instruments.

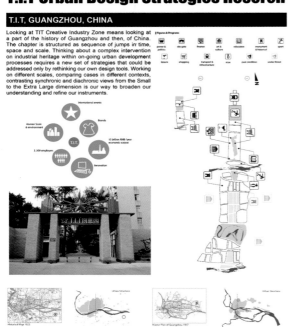

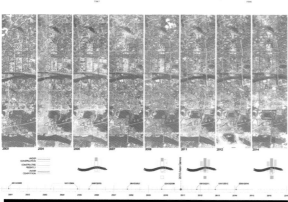

RESERCH PROCESS

Urban design can be regarded as a multifaceted concept which includes socio-economic, ecological, technical, political and ethical perspectives. Decision problems in the domain of urban design represent "weak" or unstructured problems since they are characterized by multiple actors, many and often conflicting values and views, a wealth of possible outcomes and high uncertainty. Under these circumstances, the evaluation of alternative scenarios is therefore a complex decision problem where different aspects need to be considered simultaneously, taking into account both technical elements, which are based on empirical observations, and non technical elements, which are based on social visions, preferences and feelings.

This research aims at proposing a multi-methodological approach for supporting strategic planning and design in the domain of urban and territorial projects. The proposed framework is based on the combined use of different tools for designing complex urban regeneration processes, following the subsequent phases for the definition of the projects (from very general transformation scenarios to a more detailed preliminary project). The multi-methodological approach is organized according to subsequent steps, involving the application of different evaluation methods, namely SWOT analysis, stakeholders analysis and Multicriteria Analysis, taking into account both the methodology of the Analytic Hierarchy Process and the Multi Attribute Value Theory.

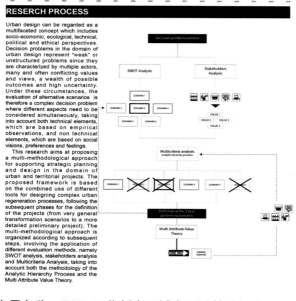

MULTICRITERIA ANALYSIS

THE PROCESS

The third phase of the process consists in the development of a Multicriteria Analysis that is a valuable and increasingly widely-used tool to aid decision-making where there is a choice to be made between competing options. It is particularly useful as a tool for sustainability assessment and urban and territorial planning, where a complex and inter-connected range of environmental, social and economic issues must be taken into consideration and where objectives are often competing, making trade-offs unavoidable.

Among the different multicriteria methods, a very important role is played by the theory of the Analytic Hierarchy Process. The AHP method allows tangible and intangible elements to be incorporated simultaneously in the evaluation, through the use of both real data and experts' subjective decisions. Following the AHP methodology, a complex problem can be divided into several sub-problems that are organized according to hierarchical levels, where each level denotes a set of criteria or attributes related to each sub-problem. The top level of the hierarchy denotes the goal of the problem and the intermediate levels denote the factors of the respective upper levels. Meanwhile, the bottom level contains the alternatives or actions considered when achieving the goal. AHP permits factors to be compared, with the importance of individual factors being relative to their effects on the problem solution, and the priority list of the considered alternatives to be reached.

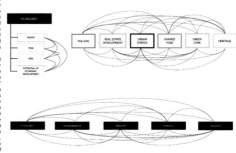

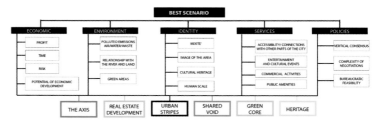

SCENARIO DESIGN

In particular, the first phase of the procedure is related to the structuring of the decision problem in order to define the goal of the project and to identify possible alternative solutions for reaching the goal. This phase is based on the development of a SWOT analysis. The acronym SWOT stands for Strengths, Weaknesses, Opportunities and Threats and the analysis is based on a logic procedure that allows the data and information on a specific decision problem to be collected and organized. With specific reference to the context of urban projects, the aim of the analysis consists in the definition of the possible development scenarios for an area, which derive from the valorisation of the strengths and the mitigation of the weaknesses, in the light of the opportunities and threats which could occur.

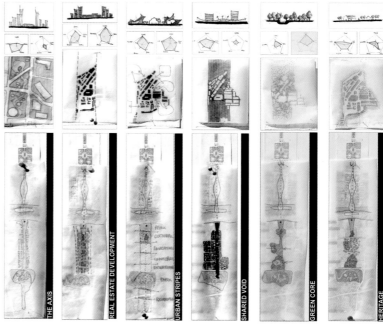

SENSITIVITY ANALYSIS

作品名称：T.I.T. 工业创意园城市设计策略探究
T.I.T. Urban Design Strategies Reserch

院校名：华南理工大学
设计人：郭晓，洪梦扬，李泳妍，倪安琪
指导教师：孙一民，Francesca Frassoldati，李敏稚
课程名称：五年级导师制
作业完成日期：2014年12月
对外交流对象：意大利都灵理工大学建筑学院

入围奖

T.I.T Urban Design Strategies Reserch

MULTI ATTRIBUTE VALUE THEORY

RE: URBAN STRIPES II

The second variation of this scenario sees as the most important value the interaction among clusters of clearly defined functions, such as cultural, commercial, entertainment and residential zones, hence the name "Urban Stripes". Different zones are divided mainly by transversal mobility, and they are highly permeable. Together they form a set of well-equipped urban machines, but they are relatively uniform within each "stripe".
Around the T.I.T. Industrial Park are four museums under construction; therefore it becomes the core of the cultural "stripe", whereas the role of T.I.T. remains very crucial: an interstitial space connecting these museums as well as a characteristic intro of the whole series of the "urban stripes".

RE: THE AXIS II

The second variation of this solution again retraces the Government Plan for the design of an urban axis to be the continuation of the new northern Axis. The proposal stresses the axial distribution of functions along a development that connects the river to the new basin lake and again to the river to the south.
This design offers the possibility to make a new strong mark on the city in its most developing district.

RE: GREEN CORE II

The second "Green Core" scenario is again intended to create a sort of urban forest, in which the environmental effect is considered as most important, not only in the dimension of ecology, but also regarding the essential quality of urban life. Therefore, the prosperous vegetation inside T.I.T. is of highly integrated value.
Following the green areas under the Canton Tower, patches of small woods cover the land and the visual axis is thus softened, winding its way through this "forests". This curve-shaped green area extends all way long until it meets the villages. "Forests" turned into aligned trees with full respect of the original borders and forms of these villages. On the other hand, high-rise buildings are arranged on the edge of the green area to maintain a certain density as well as a strong transversal permeability.

THE PROCESS

These three scenarios have been further investigated by means of the Multi Attribute Value Theory in order to better differentiate their performance and to obtain more stable results. The intention of MAVT is to construct a means of associating a real number with each alternative, in order to produce a preference order on the alternatives consistent with the Decision Maker value judgments. To do this, MAVT assumes that in every decision problem a real value function exists that represents the preferences of the Decision Maker. This function is used to transform the attributes of each alternative option into one single value. The alternative with the best value is then pointed out as the best.
From the methodological point of view, the process to be followed to build a MAVT model consists of the following five fundamental steps:
1. defining and structuring the fundamental objectives and related attributes;
2. identifying alternative options;
3. assessing the scores for each alternative in terms of each criterion;
4. modeling preferences and value trade-offs: elicitation of value functions associated with objectives and attributes and assessment of their weights;
5. ranking of the alternatives: a total score is calculated for each alternative by applying a value function to all criteria's scores
It is also interesting to notice that the proposed approach is structured according to an iterative process as it was delineated by a series of tasks, issues and feedback loops that have formed and influences the design projects during the evaluation.
Finally, the discussion oriented toward values allowed to acquire more insight into the decision, to focus on the key aspects of the problem and to drive the thinking process in an organized way. In fact, conventional approaches to decision making focus on alternatives. It has been noticed that focusing on alternatives is a limited way to think through decision situations as it is reactive and not pro-active approach. The method that has been followed in the present study was instead based on the so-called "values-focused thinking approach" that allows for being pro-active and for the creation of new alternatives and opportunities

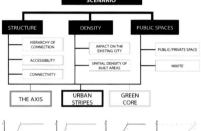

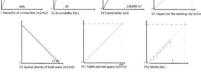

FINAL SCENARIO

RE.RE: URBAN STRIPES III

This final scenario is an attempt at converging the points of strength of the previous three scenario into one design, while minimizing their weaknesses: from the Urban Stripes design, the structure of functional clusters is kept and maximized; in particular, the museum district is organized around the T.I.T. area which is kept as intact as possible. From the Green Core scenario, an urban forest runs through the site and offers a true lung of vegetation inside the city. From The Axis scenario, the longitudinal axis remains, which connects the river to the lake and then again to the river. Additionally, a system of transversal connections allows the site to be extremely permeable to the city on the east and on the west.
Evidently, this is not an in-depth design, but rather a suggestion on how a different approach of decision making processes can influence the design of cities.

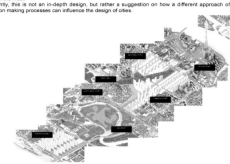

SENSITIVITY ANALYSIS

TABLE FOR MAVT ANALYSIS

MAVT SENSITIVITY ANALYSES

University: South China University of Technology
Designer: Guo Xiao, Hong Mengyang, Li Yongyan, Ni Anqi
Tutor: Sun Yimin, Francesca Frassololati, Li Minzhi
Course Name: SCUT - POLITO Joint Design Studio
Finished Time: Dec., 2014
Exchange Institute: Politecnico Di Torino University

Honorable Mentian

co-housing 1

活力台北·青年文创SOHO集合住宅设计

区位分析

基地分析

基地周边景观流线分析　基地周边交通分析　基地周边活力点分析　基地中树的关系

元素提取

日式住宅之间间隔小，形成巷道的比例尺度　　不规则形状的平面错落形成不同共享庭院

SOHO族人群研究

榻榻米元素的研究

音乐类　文字类　IT类

艺术类　手工类

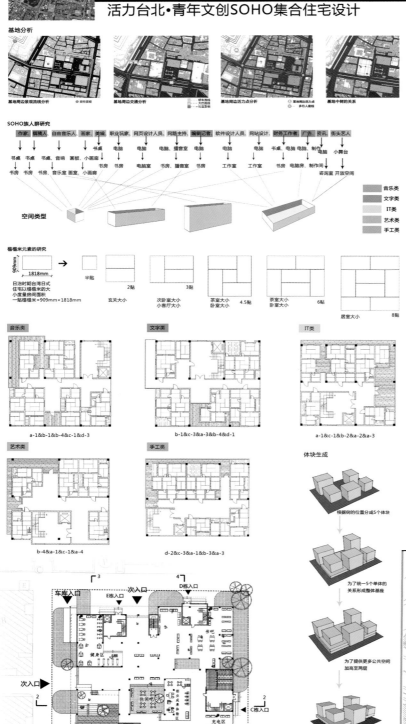

一层平面图 1:350

户型组合

1人　a-1　a-2　a-3　a-4

2-4人　b-1　b-2　b-3　b-4

5-7人　c-1　c-2　c-3

8-10人　d-1　d-2　d-3　d-4　d-5

体块生成

根据树的位置分成5个体块

为了统一5个单体的关系形成整体基座

为了提供更多公共空间加高至两层

架空一层分私密性

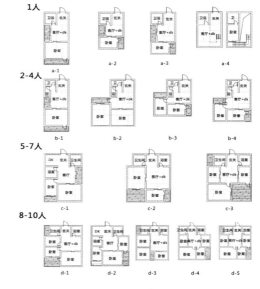

经济技术指标
占地面积：2861m²
建筑面积：9662.5m²
容积率：0.3
绿化率：22.5%
覆盖率：86.1%

总平面图 1:1000

设计说明：
本次设计基地位于台北市中正纪念堂东北角，位于华光社区中，毗邻前台北监狱t遗存的日式官舍。在长期规划中该片区域将以商办观光旅店和历史遗存青创作为概念构想，以中正纪念堂、台北监狱旧城墙、金华路沿街保留的日式住宅形成景观轴。
本设计以规划中青创商办为主要功能要求，结合现代青年自主创业的发展趋势，提取基地中日本住宅的各类元素，形成以青年创业SOHO为主体，兼具满足周边居民文娱活动的共享合作式集合住宅。
本设计以"co-housing"作为概念，即共居的理念。既使得不同类型创业人员共居交流，也使居民与创业人员、居民之间产生共居、共享的行为。

作品名称：活力台北——青年文创SOHO集合住宅设计　Co-housing

入围奖

院校名：华侨大学
设计人：王欣远
指导教师：胡璟，连旭，吴少峰
课程名称：台北华光片区青年旅馆共题设计
作业完成日期：2015年05月
对外交流对象：中国文化大学

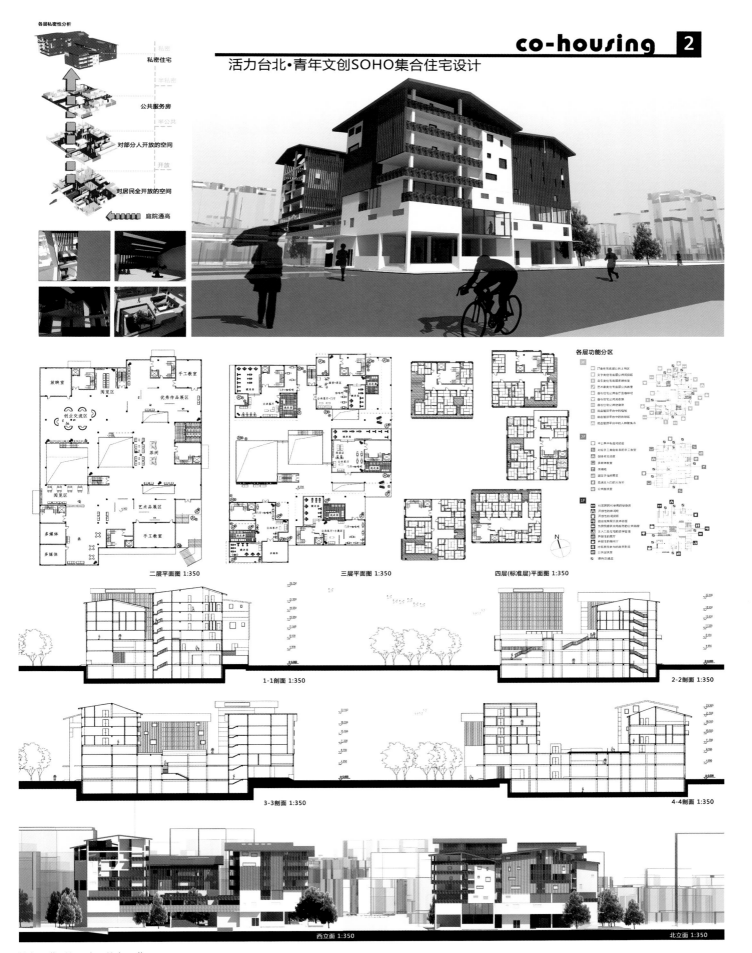

University: Huaqiao University
Designer: Wang Xinyuan
Tutor: Hu Jin, Lian Xu, Wu Shaofeng
Course Name: 3rd-Year Studio Design Assignment
Finished Time: May, 2015
Exchange Institute: Chinese Culture University

Honorable Mentian

點·聚生活 — 澳門十月初五街片區活化計劃 01

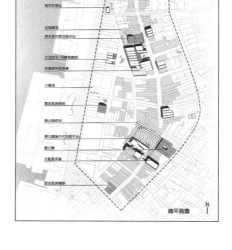

片區歷史

1553年 | 1910年10月5日 | 1920-1930年代 | 1930年代 | 1980年代 | 1990年代 | NOW—>

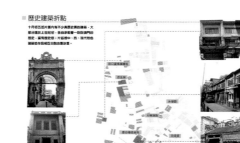

作品名稱：点·聚生活
Attitude in Our Culture Area

院校名：华侨大学
设计人：叶子颖
指导教师：费迎庆
课程名称：中国澳门十月初五街改造与更新设计
作业完成日期：2015年06月
对外交流对象：中国澳门土地工务运输局，澳门文化局

入围奖

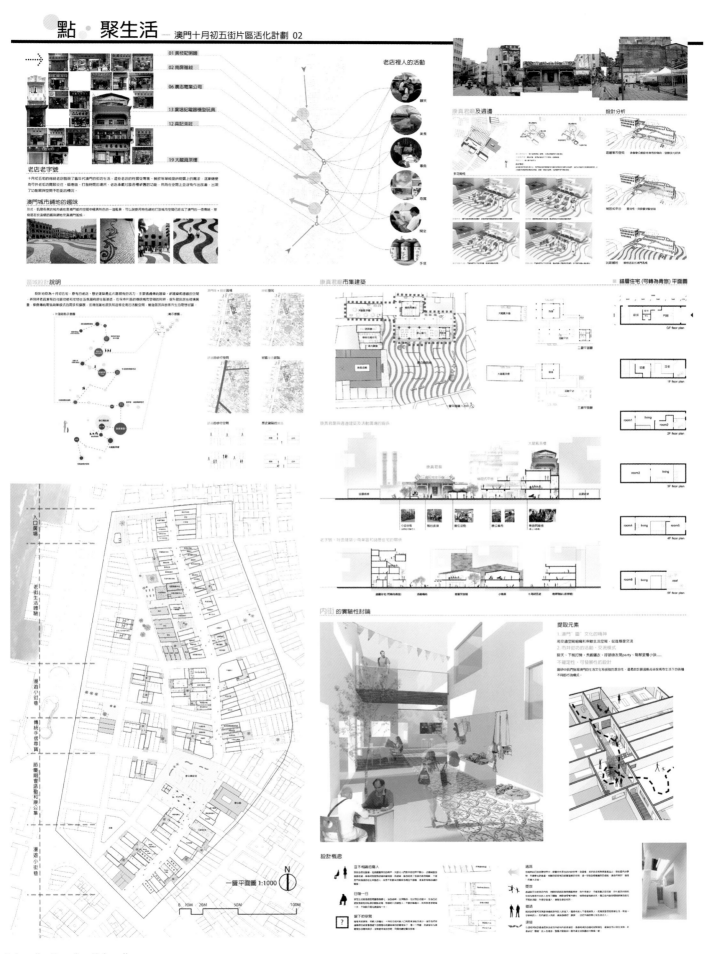

University: Huaqiao University
Designer: Ye Ziyin
Tutor: Fei Yinqin
Course Name: Redevelopment Master plan of Rua de Cinco de Outubro
Finished Time: Jun., 2015
Exchange Institute: the Land, Public Works and Transport Bureau of the Macao, Cultural Institute of Macao

Honorable Mentian

Rehabilitation Neighbourhood
For the community center

REACTIVATION

This project is a community center in a rehabilitation neighbourhood center. The rehabilitation neighbourhood center is a rebuilding project of "GuangFangLian". It includes six kinds of disease: action impairment, hearing loss, autism, depression, addiction, drug abuse. Through attract the "YuanCun" residents into the community center, we want to rebuild the life network of the rehabilitation neighbourhood center and help the patients rehabilitation program which is benefit for the intergration of new and old community.

GuangFangLian is a imprint for the process of industrialization development history, witnessed the formation of industrial landscape, the irreplaceable urban characteristics.

The community around GuangFangLian lack of centralized public activity space. A large number of residents gathered at the roadside for recreational activities.

How to properly handle the relationship between them is an important issue for the renovation project.

The site is located in a junction point of the YuanCun community and rehabilitation center, as a buffer zone for flow intersection of the old and new community. The recovery brings light on the fundamental theme of "reactivation." Reserve the existing factory building struction. Put into the appropriate function. Create a pleasant scale through the walls and skip floors. Three series of the atrium blur the boundaries of indoor and outdoor, and provide opportunities for the person flow more comfortable encounter. Natural light floods the space through its skylights. The reserved factory building struction evoke the memory of the industrial age. Here will become a cozy meeting place and sharing space.

SITE INFORMATION

CURRENT SITUATION OF LEISURE

The surrounding situation

Activity distribution

Activity status quo

Mission

Spatial scale analysis
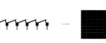   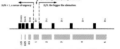

The large-span workshop' scale not suitable for social activities. | Use a module to divide the internal space. | Extract the module to combine. | Recombine to form a pleasant scale. | 0.5<D/H<1, Have a comfortable space to experience. | The witdy and height of the street form different spatial experience.

North skylight, insufficient light; Large scale; Single closed space | Divided space; Skip floor; Pleasant scale; Open the roof; Change the roof; Atrium

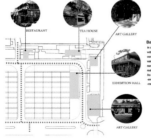

Base peripheral status quo

CONCEPT GENERATION

Reactivation

Plant analysis

Building generation process

CONNECTION | RECONSTRUCTION | DISPELLED THE MASS

STREETSCAPE | COURTYARD | THE NEW SITE

1-1 SECTION

作品名称：社区活动中心
Rehabilitation Neighbourhood : for the Community Center

入围奖

院校名：广州大学
设计人：李浩，陈彬新
指导教师：蔡凌，Ferretto Peter，邓毅，李桔，Harry Philips
课程名称：城市更新 workshop——社区康复中心
作业完成日期：2015年07月
对外交流对象：香港中文大学建筑学院，英国格拉斯哥艺术学院麦金托什建筑学院

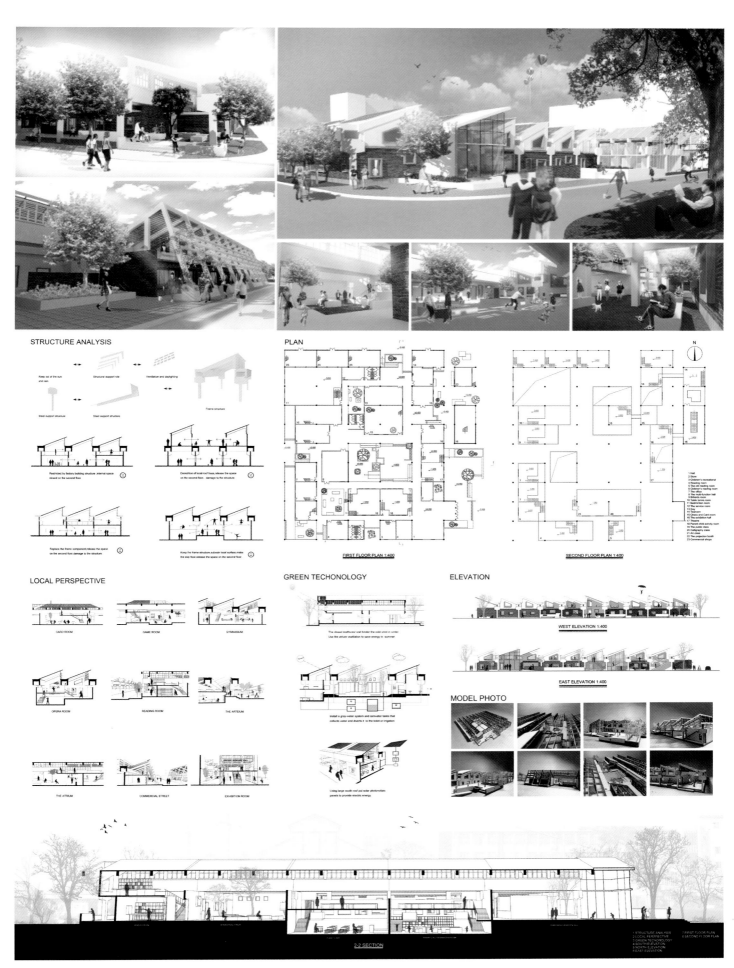

University: Guangzhou University
Designer: Li Hao, Chen Binxin
Tutor: Cai Ling, Ferretto Peter, Deng Yi, Li Ju, Harry Philips
Course Name: Workshop
Finished Time: Jul., 2015
Exchange Institute: The Chinese University of Hong Kong, The Glasgow School of Art

Honorable Mention

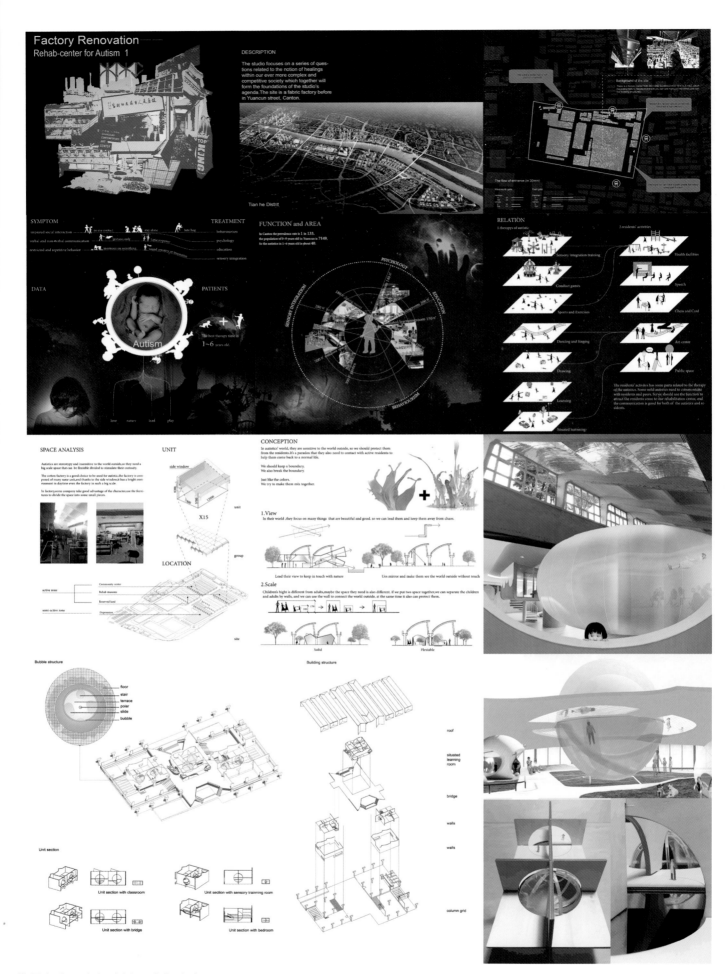

University: Guangzhou University
Designer: Hu Bin, Huang Zhishang
Tutor: Cai Ling, Ferretto Peter, Deng Yi, Li Ju, Harry Philips
Course Name: Workshop
Finished Time: Jul., 2015
Exchange Institute: The Chinese University of Hong Kong, The Glasgow School of Art

Honorable Mention

A ROUTE TO HARMONY THROUGH MARKET

In Shendun village

Village should not only be defined by its population size, or the environment characteristics. On the contrary, cultural is its fundamental meaning. Its streets and the platform is a place of comedic life show. We tried to throw a stone into the village, to inspire the natives to enjoy their daily life with our "dramatic" stage in a variety of platforms.

A broken mirror

This is a process that a mirror was hit by a stone. And it is also a process of natural variation. According to a point of view of a French architectural theorists, Abbe Laugier. He said we can find the principle of architecture from nature. In fact, every single building will follow its primitive general design. So we hope we can bring an accident into our design and this "accident" can be used to achieve our target. Those unpredictable things make this normal mirror becomes colorful and complex.

We choose a process of a mirror was hit by a stone to research. And some interesting things have been found. In this process, that the "0" stand for an accident and the other ordinary numbersrt stand for some things happened after the "0" appeared. Things will merge or split in the process of the development of production. As a result, "4.1"/"4.2" and "6(7)" stand for the division and merge of things. At the same time, those new things still appear. Exciting thing is that a complete debris produced. A fissure extending gradually and each part independent gradually. A complete mirror turn into a broken mirror because of the stone. 0 and X associate with each other, each one will not appear alone. The final result is a fantastic thing, and in other words, this rule can be applied to the building design. So maybe the "0" is a kind of way that we will change people's life. Meanwhile the number like "1"/"2"/"3" and so on, they are something we cannot predict. "X" is a kind of thing that it can be self-renewal. All of those things are the most important elements for our design.

  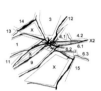

0: An accident
1/2/3/4/5/6/7/8/9/10: Something that has happened
11/12: New things
4.1/4.2: Division of the object
6(7): Merger of the object
X: Final shape
A fissure extending gradually

Each part independent gradually and pieces begin to produce
A complete mirror turn into a broken mirror
0: An accident X: Debris
0 and X associate with each other
The final result is a more perfect thing

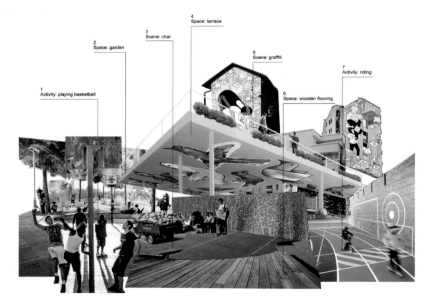

Image: the collage for Shendun Village in the future

Once the "accident" is implanted into this community, we think there are maybe some interesting things. Natives and outsiders will continue sit at the entrance of the village and chat together. Young man and other people will want to enjoy those new things and activities, such as new park, roof garden, graffiti. Those different elements can be classified for three main groups: SPACE, SCENE, ACTIVITY. And those things are the numbers in that broken mirror. So when we use a stone(0/accident) to hit a mirror(old village), we can get an amazing result(1/2/3/X/...) what we call it collage. The collage is that what we want to get.

作品名称：城中村中的新市场
A Route to Harmony through Market

院校名：广州大学
设计人：刘健，胡彬，覃可妍，劳佩珊
指导教师：邓毅，Alan Hooper，李桔，骆尔提
课程名称：冼墩村更新城市设计 workshop
作业完成日期：2015 年 05 月
对外交流对象：英国格拉斯哥艺术学院麦金托什建筑学院

A ROUTE TO HARMONY THROUGH MARKET

Site

1. Architectural Texture
2. 3-5 floors building
3. According to six ancestral halls, we draw a circle.

4. The red cross stand for road intersection.
5. We use lines to connect the red cross to complete the road system.
6. To avoid two public spaces being too close, the three positions were selected. And in the short-term, we only chose one place to implant strategy.

On the basis of the survey results, there're only a few stores(the red buildings) in the village. The villagers usually have to walk to the market(the green buildings) in Zhongcun village to buy food. We interviewed some of the villagers. Most of them still want to have their own market in the village to bring them convenience.

Accident

In the village, the entrance of the buildings sometimes is set in different orientation. That led to a strange phenomenon. Although two families live nearby, they seldom meet or chat with each other.

Based on that, we make a little change(add an accident). We put the entrances to the second floor and create a platform. In this way, when people go home, they will meet their neighbours first, then they chat, they laugh, and they have fun!

The platform combine unit 1 and unit 2 and it also provide an open space for people's activity.

The roofs of each group are connected with stairs which enrich the spatial level.

Trees and green plants are added in the communication space to make the neighborhoods alive.

Implant market

In the village, most of the ground floors in the buildings is used as a garage or storage. In order to play the value of these ground spaces and make better use of them, we re-plan the bottom space of the buildings and implant commercial functions of it. In the meanwhile, to improve the living quality of villagers, we enhance the living space into the air and create a sky garden. In this way, the activity spaces in the village become diversified and people's life will be more colorful.

A popular social activity: group dancing in an open plaza
Ballroom Dancing
Sitting on the parterre with their cushion
Chat
Sunbathe
Take exercise
Running
Riding a bicycle
Come and sing together
Sword dance
Fan dance
Act in an opera
Kicking shuttlecock
Walking the dog
Playing chess
Tell old legends or stories
Reading
Calligraphy creation
Wall graffiti art
Take a rest
Just passing by
Watching

The upper space
Natives' daily activities will mostly gather on the platform. Through this home platform, the probability of encounters between people has increased. They start various activity. Children, the elder and young people can have fun in an open space. Gradually, people in the group will become intimate.

 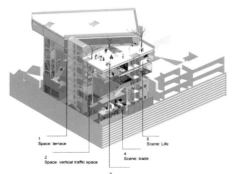

Fresh fruit stall
Meat and vegetable market
Night snack street
Spread out goods for sale in booth
Weekend fashion market
Creative handmade course
Wandering
Holiday parade
Take delivery
Play children game
Movie night
Dancing night
Art exhibition
Performance plaza
Chat in the courtyard
Enjoy the painting and calligraphy
Handmade bakery
Ice-cream booth
Second-hand goods big sale

1 Space: terrace
2 Space: vertical traffic space
3 Space: garden
4 Scene: trade
5 Scene: Life

Ground floor space
The ground floor space is re-planning. According to the type of activity, the space is divided into two regions. We set up the market in red and green regions to meet the daily needs of residents. Blue is the main entertainment area, some commercial and cultural exhibition will be hold in here. The white area is the atrium and vertical traffic space.

Three steps

Short-term: Our market will meet peoples' basic needs. Meanwhile, we add some activities(education, entertainment, culture). This is a prototype of "community center" in our network.
Mid-term: Network has been formed. The market provides OUTSIDERS and NATIVE a community with a sense of belonging and acceptance.
Long-term: VILLAGE and CITY is equal. Ultimately we achieve the goal of harmony.

University: Guang zhou University
Designer: Liu Jian, Hu Bin, Tan Keyan, Lao Peisan
Tutor: Deng Yi, Alan Hooper, Li Ju, Luo Erti
Course Name: Workshop
Finished Time: May, 2015
Exchange Institute: The Glasgow School of Art

Honorable Mentian

REBORN ETRUSCAN

Prato is a city and comune in Tuscany, Italy, the capital of the Province of Prato. The Bisenzio River, a tributary of the Arno, flows through it.

Since the late 1950s, the city has experienced significant immigration, the most notable being a large Chinese community which first arrived in the late 1980s. With more than 189,000 inhabitants, Prato is ITuscany's second largest city and the third argest in Central Italy, after Rome and Florence.

The city of Prato has the second largest Chinese immigrant population in Italy. Local authorities estimate the number of Chinese citizens living in Prato to be around 45,000, illegal immigrants included.

This project base is located in the eastern part of Italy's third largest city in prato.

Site is located in a near the ruins of the ancient Etruscan. Etruscan civilization is an important excessive Greek culture and Roman culture, which has a highly developed civilization, but it is not known to most people.

Hence, Renaissance, trulli subculture, and can make more people know about it, has become our main aim of the mission.

So we adopt the peripheral industry drives the culture in the south of the park, and built to attract tourists shopping, entertainment, so it can be able to meet people of different ages and different area.

1. museum
2. corridor
3. ruin
4. video hall
5. youth hotel
6. viewing deck
7. cookshop
8. cafe
9. shopping mall
10. restaurant
11. reception center for aged
12. retail shop
13. children's activity center
14. hand-made center
15. office

1. Fuction

The function is complex around the site, which is not museum for Etruscan and related supporting facilities.

2. Function Division

Surrounding coummunities are massive distribution, residential block for large and small industrial zone block.

3. Green

Original virescence predominantly green belts along the road, there are not many large cluster center, and landscaping.

4. Boundary

The overall direction of the city along the northwest. Site along the northwest direction followed the city toward.

5. Node
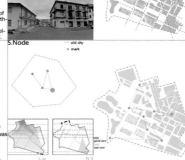
Castle is the center of the old city, the flag was the old city and the node will have some space with the castle and the sight contact.

Site Plan 1:3000

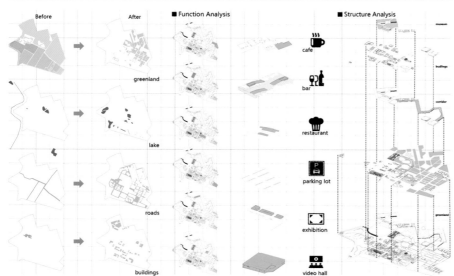

■ Function Analysis ■ Structure Analysis

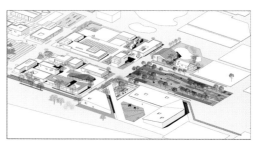

作品名称：伊特鲁里亚的复兴——普拉托Gonfient地区遗址公园设计
Reborn Etruscan

入围奖

院校名：山东建筑大学
设计人：于陈晨，张渠，张明辉
指导教师：常玮，王茹，Su Bin, Luca Piantini
课程名称：山东建筑大学—新西兰UNITEC理工学院联合设计课程教学
作业完成日期：2015年06月
对外交流对象：新西兰UNITEC理工学院

REBORN ETRUSCAN

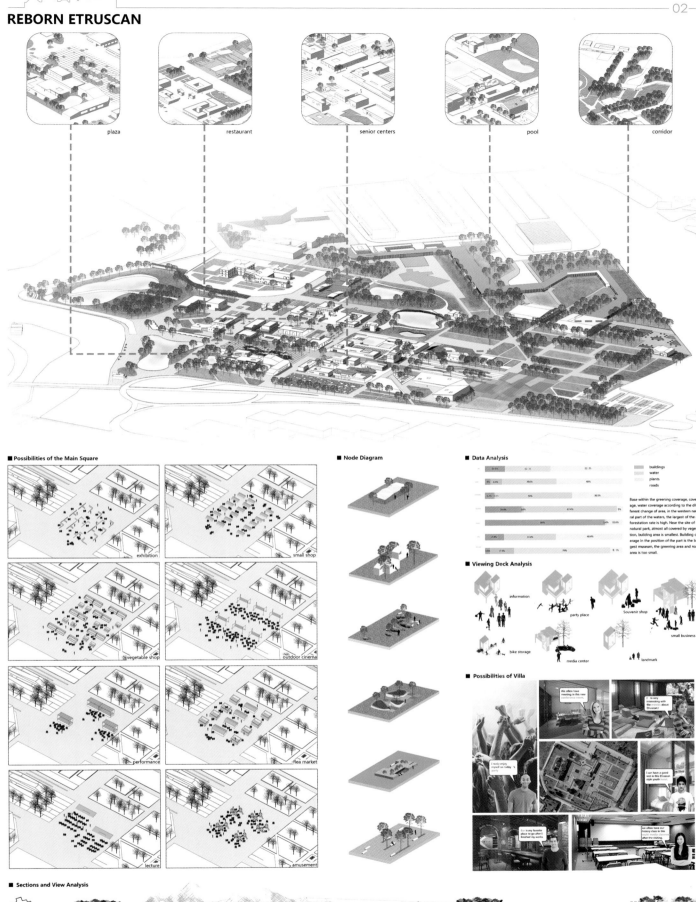

University: Shandong Jianzhu University
Designer: Yu Chenchen, Zhang Qu, Zhang Minghui
Tutor: Chang Wei, Wang Ru, Su Bin, Luca Piantini
Course Name: Shandong Jianzhu University - Unitec Institute of Technology Joint Design Studio
Finished Time: Jun., 2015
Exchange Institute: School of Architecture and Architectural Technology, Unitec Institute of Technology

Honorable Mention

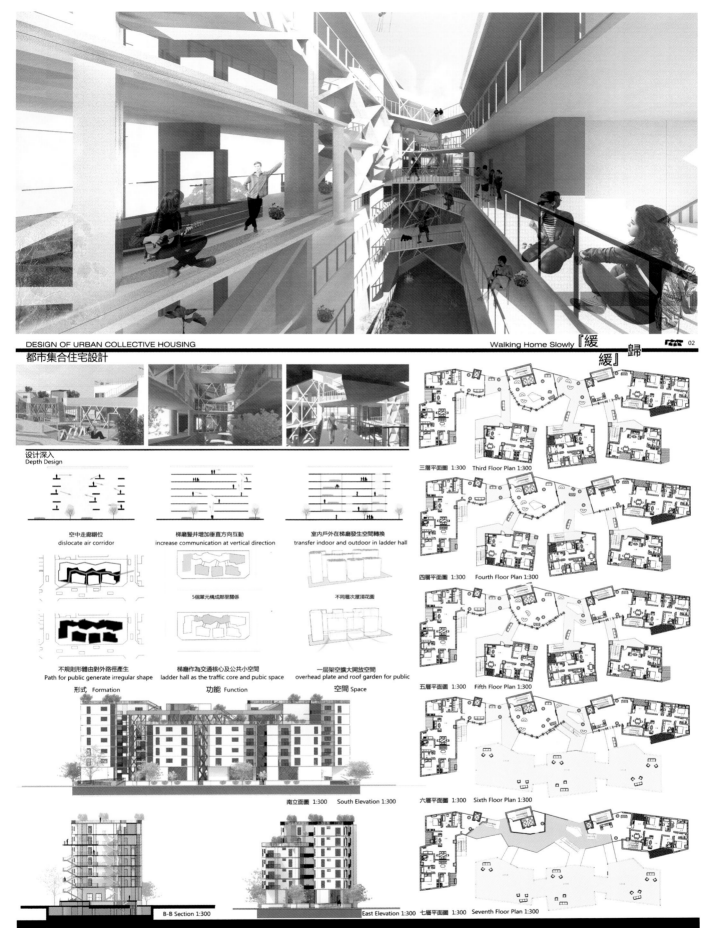

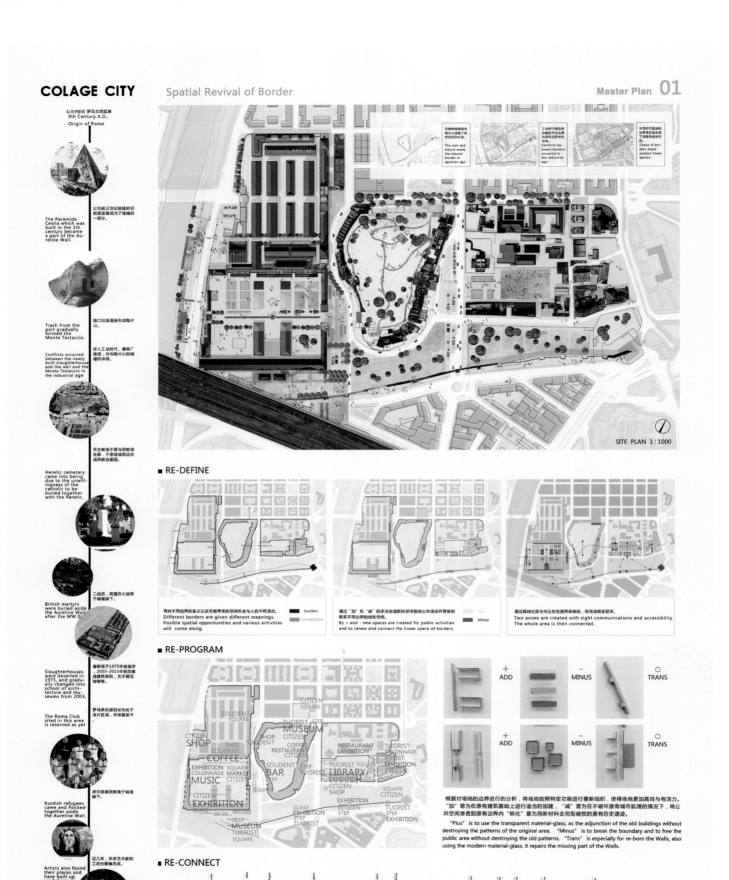

作品名称：唤醒历史——古罗马西南角城墙区更新
Wandering through the History

院校名：天津大学
设计人：李文爽，崔家瑞，李宗泽
指导教师：张昕楠，卞洪滨，赵娜冬
课程名称：古罗马西南角旧屠宰场—陶片山—城墙区域城市设计及旧建筑加改建设计
作业完成日期：2015年08月
对外交流对象：意大利罗马大学建筑学院

入围奖

COLAGE CITY Spatial Revival of Border

Architecture Design 02

Throughout the developing of Rome, new constructions don't mean ruining the old. Instead they coexist to serve a new life. Urban fabric of Rome becomes unique, and the historic marks are best reserved.

CULTURE COLLAGE IN CITY

CULTURE COLLAGE WITH DEFENSIVE WALLS

■ FACTORY PLUS

The project is in purpose to energize the area through the way of opening the abandoned space to public area. I identify this abandoned factory as a new community center, while the artists living here and can give the visitors a casual lecture

SECOND FLOOR PLAN 1:200   GROUND FLOOR PLAN 1:200

■ SLAUGHTERHOUSE PLUS

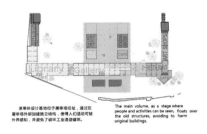

The main volume, as a stage where people and activities can be seen, floats over the old structures, avoiding to harm original buildings.

■ THE WALL PLUS

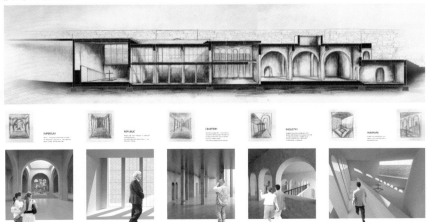

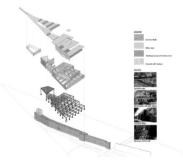

I think about how do historic objects such as complex wall systems provide a support for the redevelopment of the surrounding city, how can the modern city surrounding the walls be modified in order to accommodate contemporary urbanity, while leveraging on the intrinsic qualities of the walls in order to become more attractive?

University: Tianjin University
Designer: Li Wenshuang, Cui Jiarui, Li Zongze
Tutor: Zhang Xinnan, Bian Hongbin, Zhao Nadong,
Course Name: Urban Design in Roma for the Refreshment of the Adjacent Area of Aurelian Wall and Reconstruction Design
Finished Time: Aug., 2016
Exchange Institute: School of Architecture, Sapienza University of Rome

Honorable Mentian

VERTICAL TERMINAL
----Network-3D Tall Buildings as Extensions of Urban Infrustructure and Vitality

As one of the busiest metropolitans in the world, New York city enjoys a reputation of high-efficient transportation system. But after a series of research and site studies, we found that, as the density keeps increasing, there have been already explicit indications of **DECLINE IN THE CONNECTIVITY AND INTEGRATION** among major airports, regional railway system and subway system, which is the short slab of the whole urban transportation system of New York City.

Besides, known as an exemplified vertical city, New York City has one of the best underground infrastructure network around the world, but this underground development is based on extremely high cost since the peninsula of **MANHATTAN LIES ON THE LAYER OF HARD ROCK**. The city needs alternative development in future.

On the other hand, the existing high density of Manhattan makes it almost impossible for the infrastructures to expand horizontally.

Finally, these hard facts lead us to a strategy of **DEVELOP THE CITY INFRUSTRUCTURE UPWARD**, which is the fundamental basis of our proposal.

Our site is just besides the Grand Central Terminal, the busiest transportation hub in New York. By 2050, a number of more than three times the total population of Boston commuters will pass by the GCT everyday. In order to satisfy the huge future demand of commuting and enhancing the connectivity among the airports, subway and long distance commuting, we proposed a **VERTICAL TERMINAL**.

EXISTING PROBLEMS AND CORRESPONDING STRATEGY

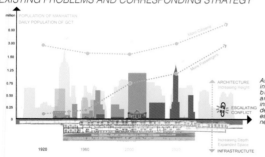
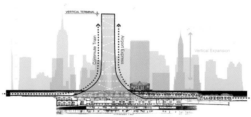

As the population grows, the increasing density pressured buildings to grow higher and higher, meanwhile city infrastructure goes deeper and deeper below the ground. The escalating conflict in-between needs a way out.

The VERTICAL TERMINAL converts the horizontal extension of common infrastructure excavation into a vertical one. By pulling up the railway, it creates a multi-layer **HORIZONTAL TRANSFER**, which is a more financially reasonable solution on account of the hard rock of Manhattan islands.

CONNECTIVITY AND INTEGRATION

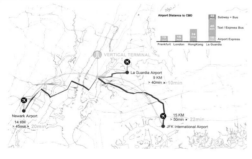
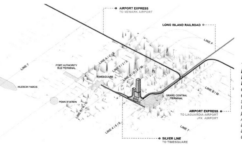

Besides the Grand Central Terminal, this site has a natural geography advantage of being at the center of 3 main airports. This compensates the short slab of NY's poor link between airports and public transportation.

The Vertical terminal **INTEGRATES AIRPORT EXPRESS TRAINS, SILVER LINE AND LIRR** into one efficient system with the existing GCT transport system, so as to maintain its function and meanwhile preserve it as a historical site. And the transfer between any two transport systems is a horizontal one within 100m distance.

SYSTEM ANALYSIS

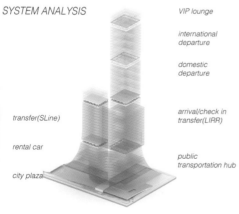

PROGRAM SYSTEM　　**STRUCTURE SYSTEM**　　**TRANSPORTATION SYSTEM**

作品名称：垂直终点站
Vertical Terminal

入围奖

院校名：同济大学
设计人：李鳌，刘晓宇，陆伊昀，杨之赟
指导教师：谢振宇，王桢栋，谭峥
课程名称：同济大学—CTBUH—KPF 联合设计课程
作业完成日期：2015 年 01 月
对外交流对象：世界高层建筑与都市人居学会（CTBUH），KPF 建筑设计事务所

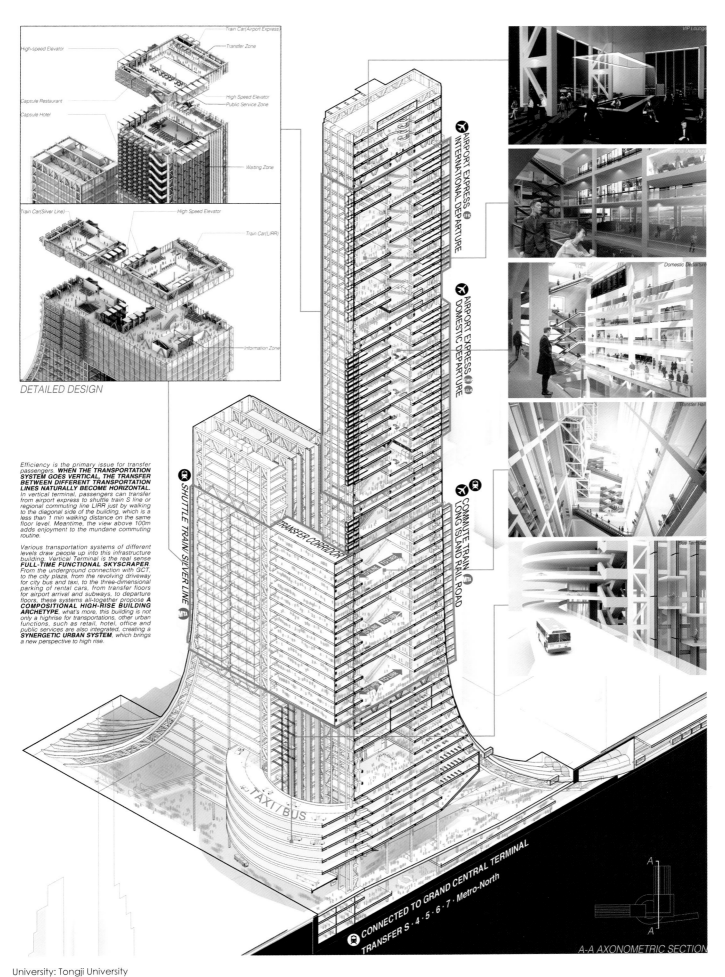

University: Tongji University
Designer: Li Ao, Liu Xiaoyu, Lu Yiyun, Lu Yiyun, Yang Zhiyun
Tutor: Xie Zhenyu, Wang Zhengdong, Tan Zheng
Course Name: Tongji - CTBUH - KPF Joint Studio
Finished Time: Jan., 2015
Exchange Institute: Concil on Tall Buildings and Urban Habitat(CTBUH), KPF

Honorable Mentian

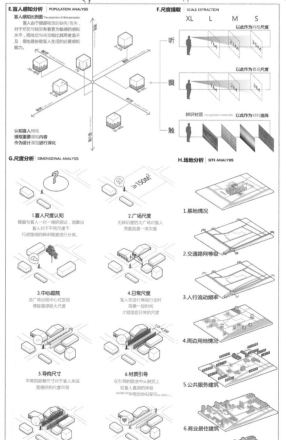

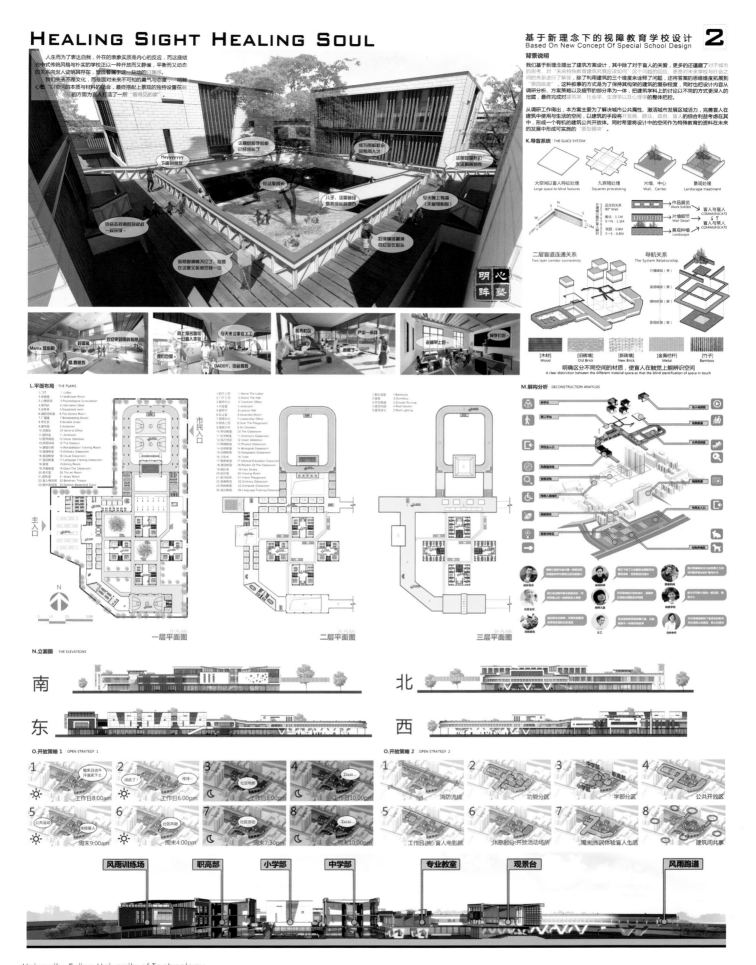

ICE CITY

Master Plan 01

Existing routes are maintained with new connections between key areas using the existing urban grid.

Pocket gardens are created as well as a green band along the old train lines which extends into the site.

A number of key features have been identified within the proposed masterplan which will inform the character of the different areas.

Low building heights in the west area of the site emphasises the church and station as a key feature.

A number of buildings are maintained to keep a sense of community whilst others are re-used for new purpose and further buildings are demolished and replaced with new.

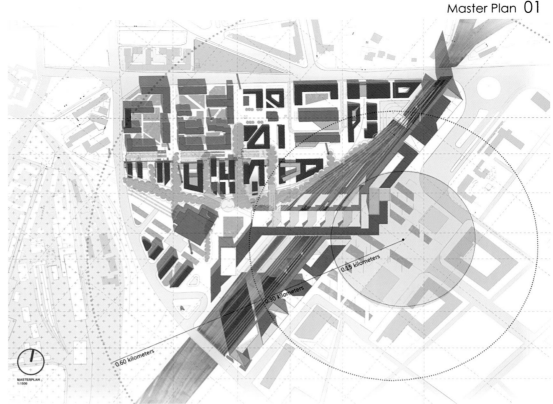

EXISTING RAILWAY YARDS

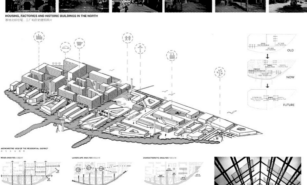

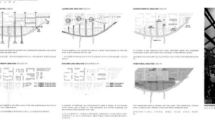

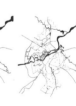

New station connecting the North and South sides of the site.

Selected buildings are retained and upgraded to maintain a sense of community.

A new market street in the residential district is created with a canopy for winter months.

The main railway repair sheds at the north entrance will be kept for use as a railway museum remembering the history of the site.

Haicheng bridge will be removed in place of a ground level road allowing access across the site.

Underground parking will be provided to the South plaza creating a pedestrian only zone with bus and taxi points underground.

Old industrial factories will be inhabited as exhibition and creative spaces as part of a creative quarter in the residential district.

THE REGENERATION OF HARBIN CENTRAL STATION

作品名称：冰城印象
Ice City

入围奖

院校名：内蒙古工业大学
设计人：梁嘉辉
指导教师：李冰心，Jonathan Richard Bassindale
课程名称：中英建筑学生工作坊——哈尔滨火车站北区域城市设计
作业完成日期：2014年09月
对外交流对象：英国曼彻斯特大学，朴茨茅斯大学，西英格兰大学

ICE CITY

Architecture Design 02

PERSPECTIVE RENDER SHOWING STATION ENTRANCE

EXISTING STATION AND PUBLIC SQUARE

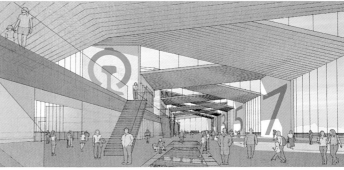
INTERIOR PERSPECTIVE OF THE TRAIN STATION WAITING AREA

The initial design concept for the individual buildings particularly the new train station and associated area is largely driven by harbin ice sculptures taking inspiration from thier jagged abstract forms which we have used to create a unique stimulating skyline across the station square.

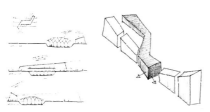

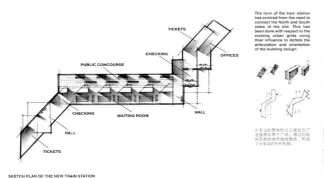
SKETCH PLAN OF THE NEW TRAIN STATION

The form of the train station has evolved from the need to connect the North and South sides of the site. This has been done with respect to the existing urban grids using thier influence to dictate the articulation and orientaiton of the building design.

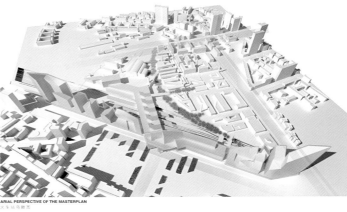
ARIAL PERSPECTIVE OF THE MASTERPLAN

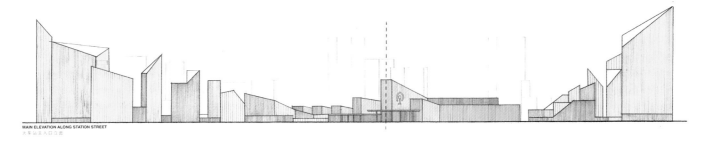
MAIN ELEVATION ALONG STATION STREET

THE REGENERATION OF HARBIN CENTRAL STATION

University: Inter Mongolia University of Technology
Designer: Liang Jiahui, Zhang Xiaoyu, Huang Na, Joshua Iain Shepherd Cherry
Tutor: Li Bingxin, Jonathan Richard Bassindale
Course Name: Sina - UK higher education collaboration on architecture design
Finished Time: Sep., 2014
Exchange Institute: University of Manchester Manchester Metropolitan University, University of portsmouth, University of the West of England

Honorable Mentian

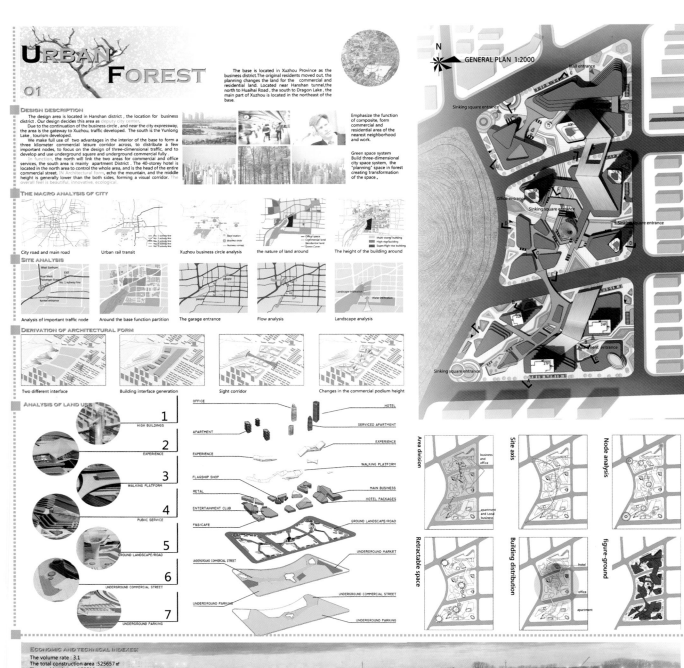

作品名称：都市森林——徐州韩山东路地块城市设计
Urban Forest

院校名：中国矿业大学
设计人：王菁蔓，齐美芝，顾贺鸣，王萌，王凯圣
指导教师：Kay Drewling，姚刚，刘茜
课程名称：中国矿业大学—奥克兰大学—联合设计课程教学
作业完成时间：2015年07月
对外交流对象：新西兰奥克兰大学建筑与规划学院

入围奖

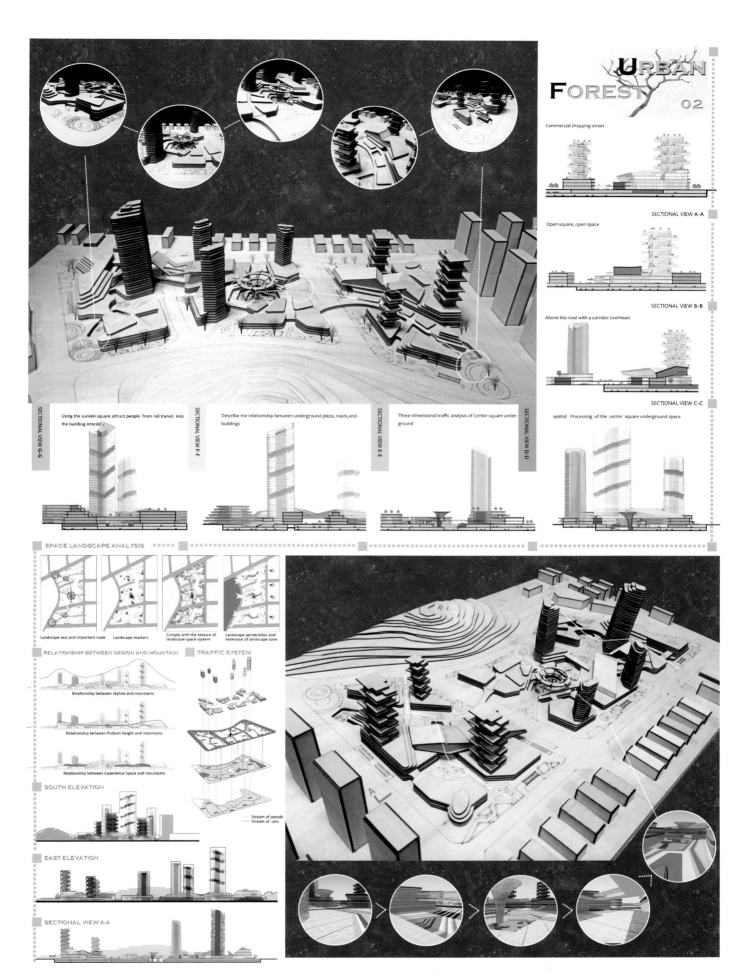

University: China University of Mining Technology
Designer: Wang Jingman, Qi Meizhi, Gu Heming, Wang Meng, Wang Kaisheng
Tutor: Kay Drewling, Yao Gang, Liu Qian
Course Name: China University of Mining Technology- University of Auckland, School of Architecture and Planning Joint Design Studio
Finished Time: Jul., 2015
Exchange Institute: University of Auckland, School of Architecture and Planning

Honorable Mentian

RURAL SERVICE CENTER

Location: Henglu, Jiangxi, China
Type: Service Building
Area: 1560㎡

Background

After a rapid development, Henglu village, a impoverished village, becomes rich, local people demolish their old houses and rebuild new ones instead. In this village, modern and traditional dwellings are mixed used. Moreover, young people work outside the village for the purpose of higher income, their old parents and children are left in this village. This situation reflect several problems, such as the shortage of local labors, education and pension problems and so on.

Therefore, this service station design focuses on the social problems, aiming at creating a place where pay more attention to the old people and children, and arousing public concerns.

New houses are build along the road

Old dwellings co-exist with new ones

Use Scenes

The main crowd in the village is the aged people and children during weekends students from art college which is 10km far will serve as volunteers in this station.

From Monday to Friday, aged people could drink tea in tea-house, play Mahjong or stay with neighborhoods viewing the lake in lounge bridge. In the evening, children come back from school, they can review and discuss their homework in book bar or watch movies with their grandparents. On weekends the students from schools of art come here as volunteers to give a vivid art lesson for aged people and children.

Throughout the festival, the young people come back from outside cities, the space will become the reunion part as family time.

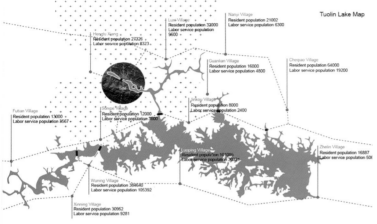

Resources

Many villages are located around Tuolin Lake, they share the rich forest and water resources. Wood, such as pine and bamboo, are used as building materials, and wood exportation is the main village revenue. Also, fishery is important to the locals, but now there are less and less people engaged in fish farming.

Forest coverage rate up to 70%

The annual average precipitation is 1437mm

Forest Resources / Water Resources

Form and Nature

We try to pull all functions into a spindly strip volumn as it would minimally interfere the natural environment. Adapt to the landform, tea room, video room and bridge are independently set up. Meanwhile, the gray space make connection with different parts of function rooms. Also, two natural open fields by using natural sloping fields are created to make space more abundant and interesting.

Land Area: 19041.5㎡
Built-up Area: 1560㎡
Building Density: 8.2%
FAR: 0.08
Greening Rate: 82.7%

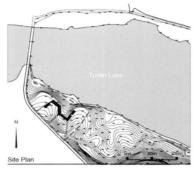
Site Plan

Form Generation

Messy Function → Function Recombination → Crossing Corridors → Adapt to Terrain

Activity Cycle and Frequency (Aged People, Children and Volunteer)

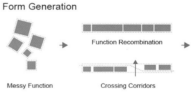

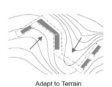

Social Problem

Migrant workers are engaged in the decoration industry, labor export has now become a main way to gain their income. However, a population of only 20,000 townships, there are 8,000 people working outside, stay for the old poeple and children, and many practical problems are constantly highlighted. Compared to rapid development of econmic income, cultural activity is deficient. How to remit the unbalanced status and attract general attention from social for the similar problems have become the key point in this design project.

Labors work outside the village

Old People

Children

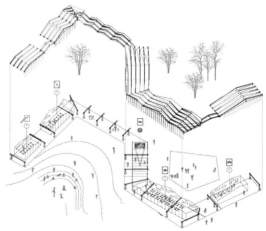

作品名称：农村社区中心
Rural Service Center

入围奖

院校名：河南工业大学
设计人：殷静怡
指导教师：王祖远
课程名称：建筑设计
作业完成日期：2015年05月
对外交流对象：澳大利亚莫纳什大学

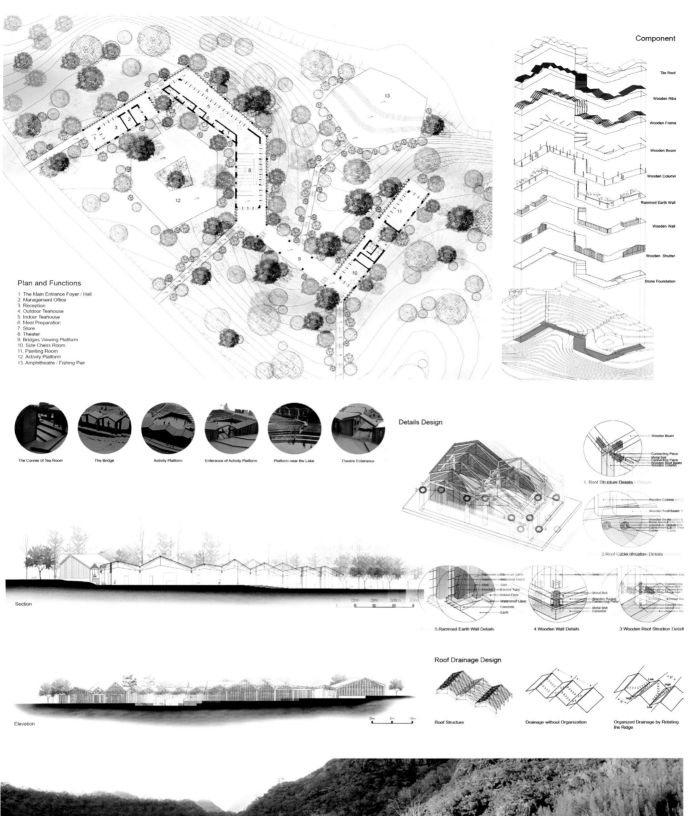

University: Henan University of Technology
Designer: Jingyi Yin
Tutor: Wang Zuyuan
Course Name: Joint Design Studio
Finished Time: May., 2015
Exchange Institute: Monash University

Honorable Mentian

LIVING + VISITING RESIDENTIAL BUILDING 1
—— Under City, Above Mount

居住+观光混合居住区设计——城市之下，山丘之上

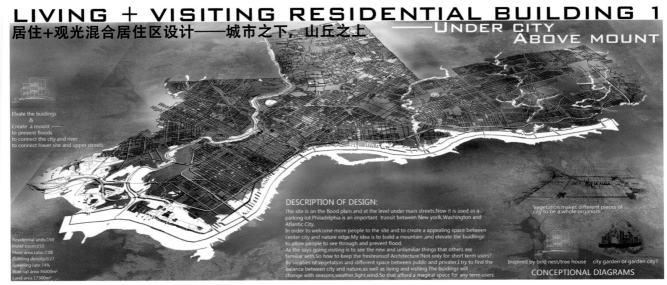

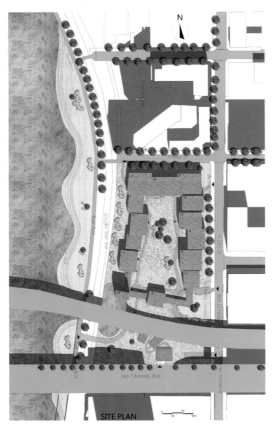

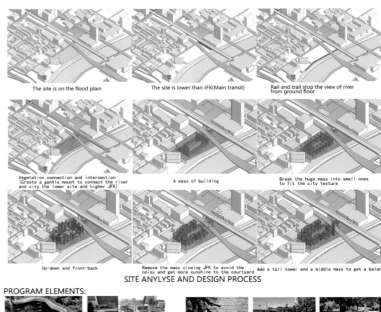

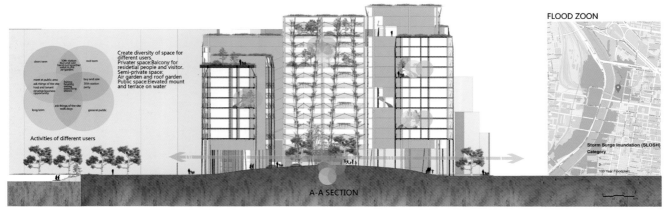

作品名称：城市之下，山丘之上
　　　　——居住+观光混合居住区设计
Under City, Above Mount : Living+Visiting Residential Building

入围奖

院校名：厦门大学
设计人：陈文倩
指导教师：Sally Harrison，张燕来
课程名称：混合居住区设计
作业完成日期：2015年08月
对外交流对象：美国天普大学建筑学院

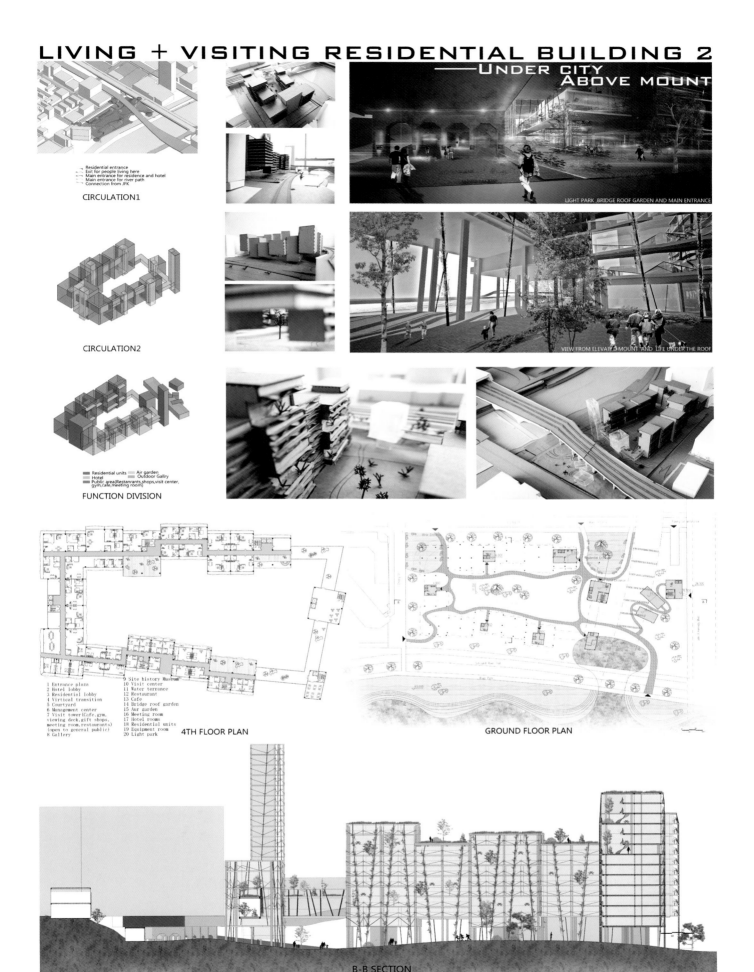

CONNECTED CITY & VIBRANT HARBOUR: The Rejuvenation Design of Jurong Industrial District, Singapore

01

REGIONAL BACKGROUND

SINGAPORE HAS BECOME ONE OF THE MOST DEVELOPED COUNTRY AFTER 50 YEARS DEVELOPMENT. THIS RELIES ON THE FULLY USE OF THE SINAPORE STRAIT AND JURONG HARBOR. JURONG IS SINGAPORE'S FIRST AND THE MOST IMPORTANT INDURY DISTRICT. ITS WHICH IS VERY CONNECTED TO THE COUNTRY'S HISTORY.

JURONG'S MAIN INDUSTRIED INCLUDE SHIPBUILDING, IRON AND STEEL, PETROCHEMICAL AND ELECTRONIC ETC.. JURONG, TOGETHER WITH TUAS AND JURONG ISLAND CONSTITUTE THE LARGEST INDUSTRIAL CLUSTER OF THE MALACCA AREA. OUR DESIGN RANGE IS AS FOLLOW.

MASTER PLAN

EXISTING SITUATIONS
CURRENT LAND USE MAP: HYBRID FUNCTIONAL ZONING

MORE THAN 10 KINDS OF INDUSTRAIL CATEGORIES LACK OF CO-ORDINATION CAUSE THE LOW EFFICIENCY OF INDUSTRIAL COLLABORATION. LARGE LOGISTICS LAND PROPORTION INCREASES TRAFFIC VOLUME. LOW EFFICIENCY TRANSPORTATION LEADS TO OVER-CROWDED STREETS.

ROAD TRAFFIC STATUS AND NODES. TWO HIGHWAYS BLOCK SOUTH-NORTH TRAFFIC WITH ONLY 4 EXITS. WE ALSO ANALYSE TRANSPORTATION STATUSBASED ON SPACE SYNTAX ANALYSIS.

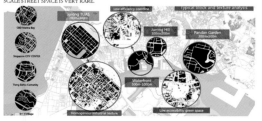

THE FUNCTION OF LOTS ARE ALMOST INDUSTRY WITH EXTREMELY LACK OF COMMERCIAL SERVICE FACILITIES AND URBAN VITALITY WITH HOMOGENEOUS AND HUGE BLOCK. HUMAN SCALE STREET SPACE IS VERY RARE.

DESIGN CONCEPT

JURONG'S INDUSTRY WILL BE INTELLIGENT DEVELOPMEN. SO WE SUGGEUST TO IMPROVE THE CONNETION OF THE INDUSTRY AREA TO INCREACE EFFICIENCY AND ALSO CREATE A VIBRANT URBAN AREA AROUND THE HARBOR BY ECOLOGY RESTORATION AND WATER CLEANING AND STORAGE. THIS IS THE CONCEPT OF OUR DESIGN, CONNECTED CITY, VIBRANT HARBOR.

CONNECTED CITY
CONNECT 3 PARTS TO FORM THE CIRCULAR TRANSPORTATION ORIENTED DEVELOPMENT CORRIDOR FOR FUTURE DEVELOPMENT
ESTABLISH THE HIGHLY EFFICIENT MOVEMENT AND TRAFFIC SYSTEM IN PEOPLE AND GOODS TO INTEGRATE THE GLOBAL, REGIONAL AND NATIONAL FUNCTIONS.

VIBRANT HARBOUR
THE DAM WOULD BE BUILT IN THE FUTURE TO FORM THE RESERVOIR BOTH FOR WATER CATCHMENT AND LIVING, WORKING, RECREATION FUNCTIONS.
PROMOTE THE RATIONAL ALLOCATION AND REUTILIZATION OF WATERFRONT FOR ACTIVITIES OTHER THAN PRODUCTION AND INCREASE THE PUBLIC ACCESSIBILITY, SPATIAL PUBLICITY AND VISIBILITY OF GREEN CHANNELS.

SPATIAL STRUCTURE

GIS FACTORS ANALYSIS

WATERFRONT EXISTING MODE
WATERFRONT PROPOSED MODE

作品名称：优联都市，活力港湾——新加坡裕廊工业区总体城市设计
Connected City & Vibrant Harbour: The Rejuvenation Design of Jurong Industrial District, Singapore

入围奖

院校名：清华大学
设计人：程思佳，曹哲静，韩靖北
指导教师：唐燕，钟舸，边兰春
课程名称：总体城市设计
作业完成日期：2015年06月
对外交流对象：新加坡国立大学设计与环境学院

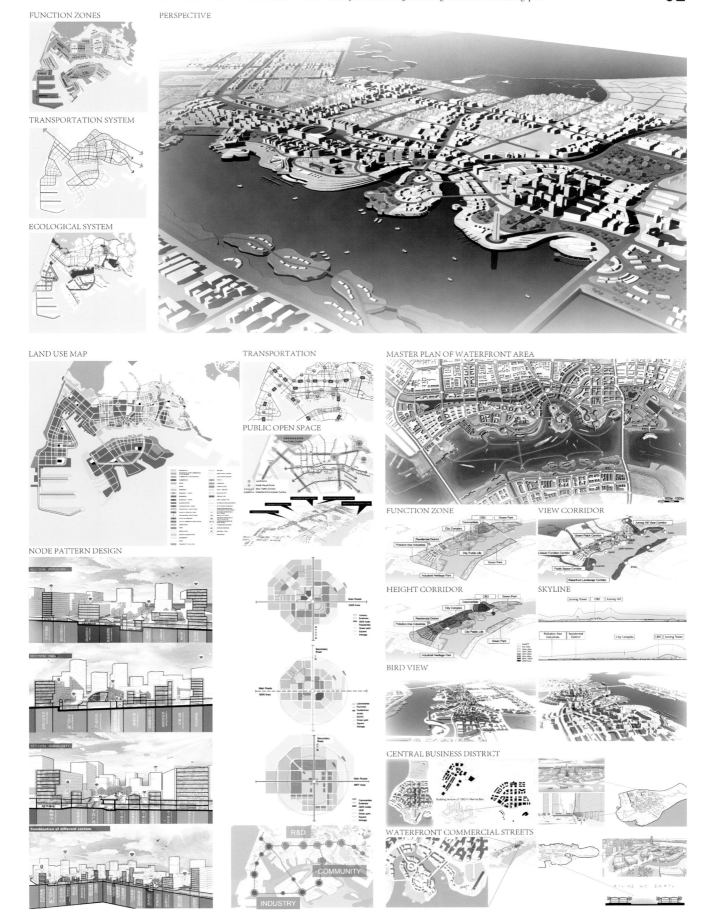

CONNECTED CITY & VIBRANT HARBOUR
The Rejuvenation Design of Jurong Industrial District, Singapore

University: Tsinghua University
Designer: Chengg Sijia, Cao Zhejing, Han Jingbei
Tutor: Tang yang, Bain Lanchun, Zhong Ge
Course name: Structural Urban Design
Finished time: Jun., 2015
Exchange institute: National University in Singapore

Honorable Mentian

W.O.W. HALL EXPANSION
THE COMMUNITY CENTER FOR THE PERFORMING ARTS

01

The W.O.W. Hall is a performing arts venue in Eugene, Oregon, United States. It was formerly a Woodmen of the World lodge, first opened at 1932. The W.O.W. Hall was listed on the National Register of Historic Places in 1996. The subject of this studio is the programming, planning and design of a new education wing to W.O.W. Hall. My basic conception is how to differentiate the new and the old in a humble way, and to activate the solid building of a long history.

SITE PLAN

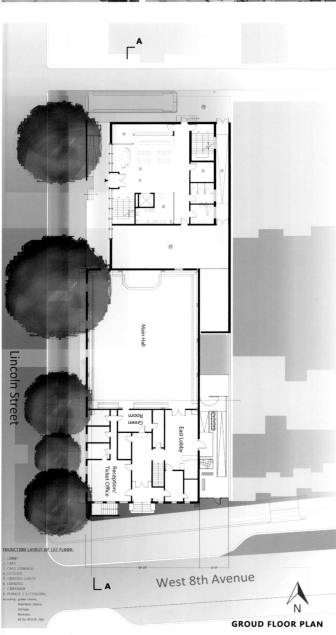

GROUND FLOOR PLAN

2ND FLOOR PLAN

3RD FLOOR PLAN

4TH FLOOR PLAN

BASEMENT PLAN

Since the west side faces a main street-Lincoln Street, it has the most interesting view and should be designed attractive enough to make more people step in. The south and the east sides are closed to the old, more private and need more consideration so as to balance the contrast between the old and the new.

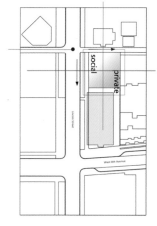

作品名称：W.O.W. 大厅扩建
An Expansion to W.O.W. Hall

院校名：清华大学
设计人：金爽
指导教师：Brian Cavanaugh
课程名称：ARCH 484/584 建筑设计
作业完成日期：2015 年 12 月
对外交流对象：美国俄勒冈大学

入围奖

W.O.W. HALL EXPANSION THE COMMUNITY CENTER FOR THE PERFORMING ARTS

EXTERIOR PERSPECTIVE VIEW

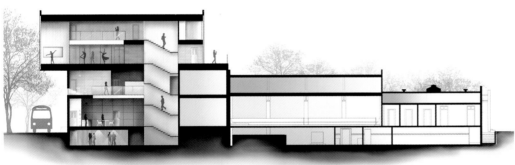
A-A SECTION WITH W.O.W. HALL

WEST ELEVATION

- The transparent glass box is served as rehearsal room to accommodate students' activities like dancing or gathering, so it will bring passers-by a lively feeling- to see and to be seen.

- And the main staircase is for lights coming into the lobby and other spaces inside the building so glass is applied. As the result when it turns dark, the light from the expansion of the W.O.W. Hall could be seen from the main street, passers-by can see what is happening in the building.

- For the entrance I choose wood because it's softer compared to outside wall.

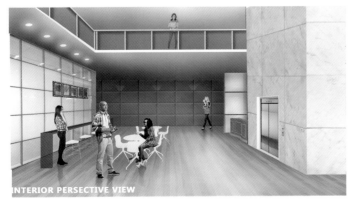
INTERIOR PERSPECTIVE VIEW

FLOW PATH CLOSED SPACE SEMI-OPEN SPACE OPEN SPACE

MORPHOGENESIS   MAIN STAIRCASE WITH GROUND GLASS.   REGULAR FIRE RESISTANT STAIRCASE & ELEVATOR

University: Tsinghua University
Designer: Jin Shuang
Tutor: Brian Cavanaugh
Course name: Arch 484/584 Architectural Design
Finished time: Dec., 2014
Exchange institute: School of Architecture, University of Oregon

Honorable Mention

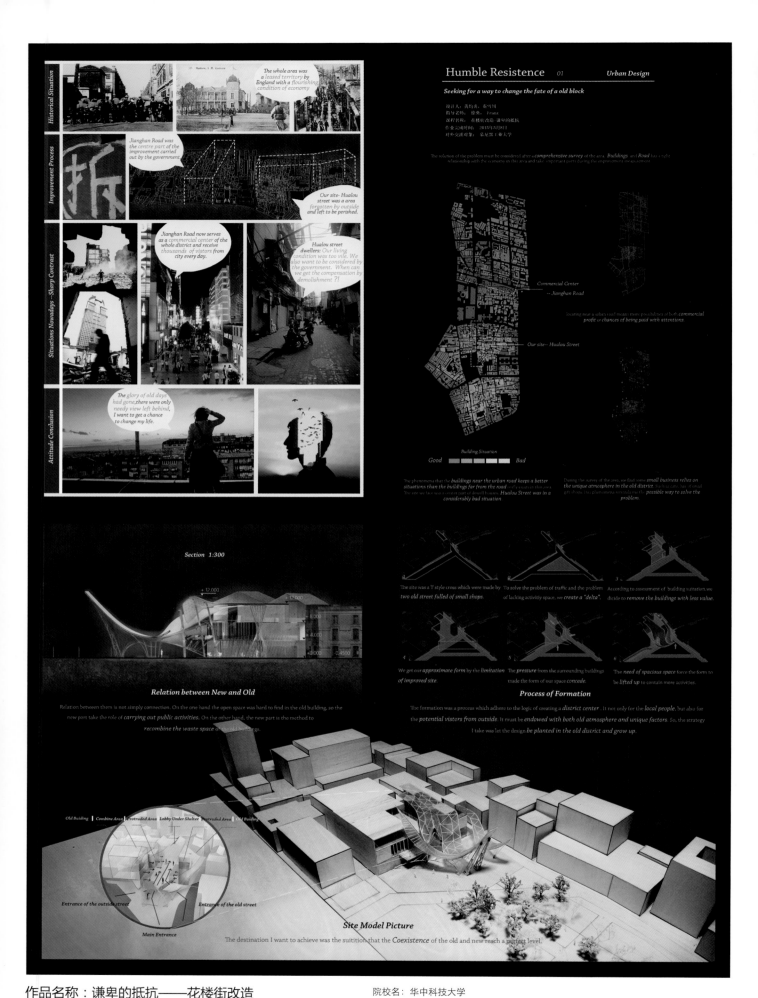

作品名称：谦卑的抵抗——花楼街改造
Humble Resistence—Hualou Street Restoration

入围奖

Humble Resistence 02   Urban Design

Seeking for a way to change the fate of a old block

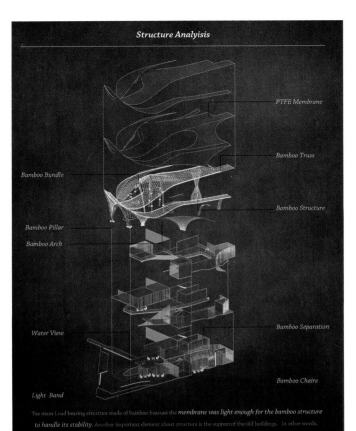

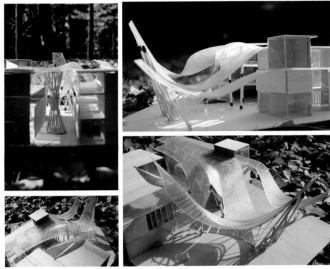

Left Above : Main Bamboo structure , View from Back
Left Below : Membrane and Bamboo Truss, View from Top
Right Above : Indoor View from Side
Right Below : View of the Entrance from Front

Using the *3D Print Technology* to present the membrane structure of the design, the model can tell not only a **strong contrast between old street and the modern facilities**, but also a **marvellous feeling which was originated by the combination of these**. The final goal which was to attract the social attention will reach only if the design made itself unique and outstanding.

Indoor Perspective Rendering

The most attractive point of the design was a *combination of the new and the old*. People can taste the retro feeling in a old city district which was unique from the model society

Old Street Elevation

Front Elevation

The elevations of the design was not made for being blended into the old district, on the contrary, I tend to *let the image prominent to the old street in order to reverse the decline of this area*

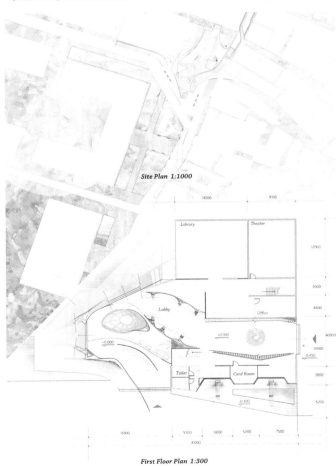

Site Plan 1:1000

First Floor Plan 1:300

University: Huazhong University of Science and Technology
Designer: Huang Junyan, Qin xuechuan
Tutor: Xu Shen
Course name: China Germany Joint Dedign Studio
Finished time: May, 2015
Exchange institute: Technische University München

Honorable Mentian

The Sinking Alley

湖底巷——武汉中心老城区复兴

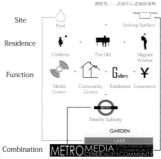

This project is aiming at **renewing the old city**, given the investigation over the issue of the area history, the land price, the selected street section and city pattern, we feel obligated to preserve all those. Therefore, **to preserve the already existing water, to increase public facilities of this area, and not to disturb the retail business already exist around the site**, we came up with a plan with architectural design, and something out of architecture.

Project Framework

As the surrounding of the site is in the center of the city, yet the value of the land is relatively higher than the adjacent area. Therefore, **the design has to be a complex of different function**

Price

Site had been characterized by its pool, which is extremely rare in the center of the city. Meanwhile, it had been well illustrated that the nearby water to the site were in a relatively long distance. In that case, **preserve of the water is not optional**.

Water

Right after the establishment of new government, the population density increased dramatically. The living condition coasted in consequence yet turning this place from the richest area in Wuhan, into the poorest place. In that case, no matter what the extra function of the site was, **it has to facilitate the local residents**.

Dwelling Situation

Origin
In order to improve the economic growth in the district, the government decide to transform a part of the residential area in the center of Hankou into commercial area in the urban planning in 1980s. One of the investor won the bid, and signed a contract with the government. As the investor acquire the tenure of the 1.6 acre, the original residents were forced to move out. As a part of the old city center, the surrounding of the site was one of the old area as mentioned.

Abandon
The construction proceeded until the investor suddenly withdraw their investment for unknown reason, just after the excavation of the foundation ditch had finished. The government encountered a dilemma, as they did not have the money to recover the land into residential area as before, nor did they accept other investor to take over the land. Yet the site had been abandoned.

Deadlock
Such situation lasted for decades while the ditch had been magically turned into a lake. Though the surrounding area became heated commercial area, no one was interested in the abandoned land. The situation awkwardly became a deadlock.

Reborn
Architect became the ice-breaker as he proposed an alternative solution. Regarding the dreadful living condition in the nearby residential area, architect persuade government invest in the re-planning of the whole area, turn the site into well-designed public place yet improving the value and the quality of the surrounding area and attracting investment to the surrounding area with the qualified public space. If the investment was successful, government just need to issue regulations to the development of surrounding area. Thus the surrounding area's renewal could be achieved without the government investments.

Brief history of the selected site

In this old district, with architectural treasure and low-wage earners, the site provided a perfect break-in for the renewal of the city as well a stage for the four roles to make the renewal happen: the **government**, the **investor**, the **residents** and the **architect**. Four roles in the program possess a unique relationship network, as they interact with each other while gaming against each other.

From **government's** perspective, their vital goal is to gain the reputation from dwellers by renewing the old district while applying the minimum expenditure. They possess the power to issue regulations and the rent the land to investors.

From **investor's** perspective, their vital goal is to achieve more profit in the low expenditure possible while co-operating with government. They are restricted by government's regulation and possess money for the lease of lands.

From the **local residents'** perspective, they are the most vulnerable. But their comment on government is of significance.

From **architect's** perspective, his vital goal is to improve the living condition of local residents while preserve the architecture treasure in the old area. In order to achieve that, he has to persuade government with a tempting plan, while attract the investor to pay for it.

To do that, government first have to pay for the renewal of the site as they turn it into a qualified public place with multiple functions. Such improvement could impose indirect influence on the value of the adjacent area and yet attract the investments to the development of these area. The money is used to achieve certain goals:

1. local residents' compensation check;
2. refine, or redesign the building under government's regulation;
3. turn the existing resident into practical use, for example, residence like it used to be, commercial utility like coffee shop;

In that case, the **architect's** goal had been achieved. The **residents** got improved living condition or huge compensation checks. The **investor** acquired the land for a lease. And the **government** gets reputation and save the money to renew the old district.

Non-architectural part of the solution

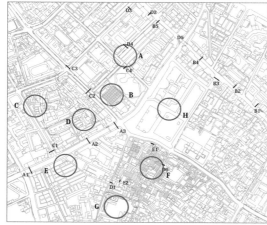

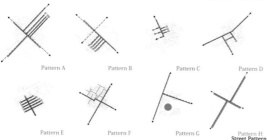

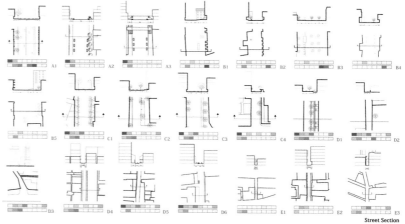

Street Section

As an expectation that our underwater complex would be designed in line with the current urban fabric, we randomly stamped 8 circle with a diameter of 100m, around the site, as a deeper investigation to the urban fabric in the site. We extracted all these samples in the circle and analyze, in order to make what we do, a salute to the old city pattern, rather than naïve imitation.

Among all those samples, there were standard traditional Wuhan residence known as Lifen, newly build commercial complex after the "Reform and Opening policy", as well the old city corner which had been irregularly constructed. We discovered that as intricate as the urban pattern might have seemed, it provided space with a scale well suited for public activity. **Meanwhile, the more complicate the urban fabric, the more diverse in the street activity. Yet, such complexity in urban fabric could be considered as a beneficial factor.**

To look further into the site, we decided to sketch of street sections. We picked up several typical part of streets for the section, analyzed its' function and activity of the street, and at last visualized it in proper scale.

Our selection of the street section distributed in five level.

A-main streets, connecting parts of the city (Cars); B-main streets, pedestrian area; C-streets, connecting main streets (Cars); D-alley, connecting streets within area (Cars); E-alley within area (No Cars).

The functions in street were classified as traffic, parking, gathering, eating, food-preparing, resting, clothes-drying and stalls.
Meanwhile, the functions in building were classified as dwelling, shops, restaurants, office, hotels, public, and vacant.

Activity Note

作品名称：湖底巷——武汉中心老城区复兴
The Sinking Alley

院校名：华中科技大学
设计人：杨隽超，王昭晅
指导教师：刘小虎，Karel Nieuwland，周钰
课程名称：中荷联合设计课程教学
作业完成日期：2015年07月
对外交流对象：Karel Nieuwland 荷兰

入围奖

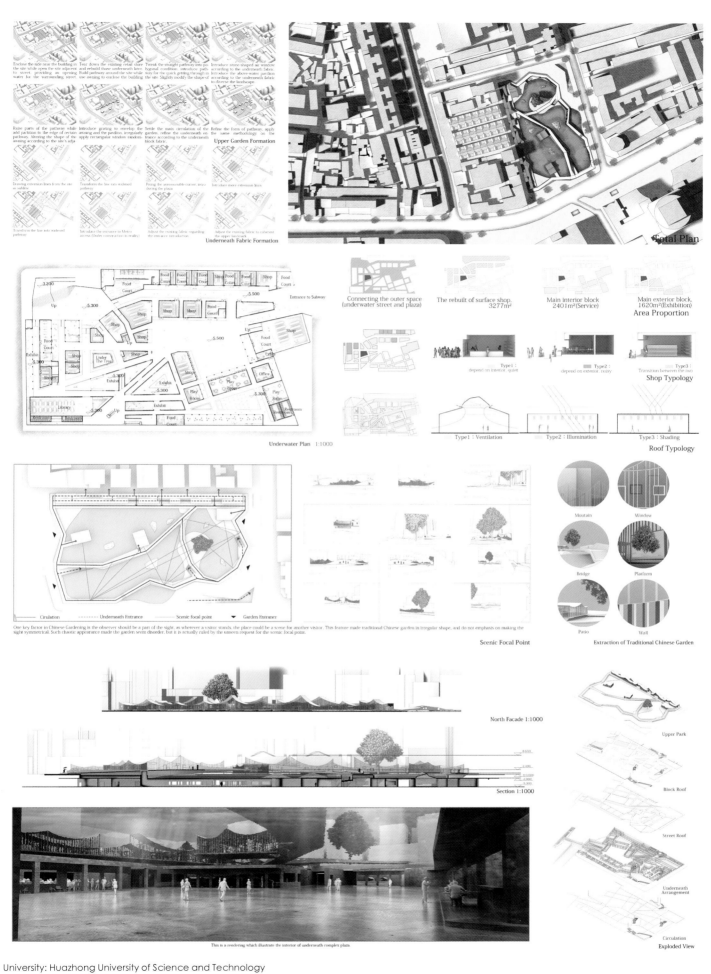

University: Huazhong University of Science and Technology
Designer: Yang Junchao, Wang Zhaoxuan
Tutor: Liu Xiaohu, Karel Nieuwland, Zhou Yu
Course name: China - Neitherlands Joint Design Studio
Finished time: Jul., 2015
Exchange institute: Karel Nieuwland

Honorable Mentian

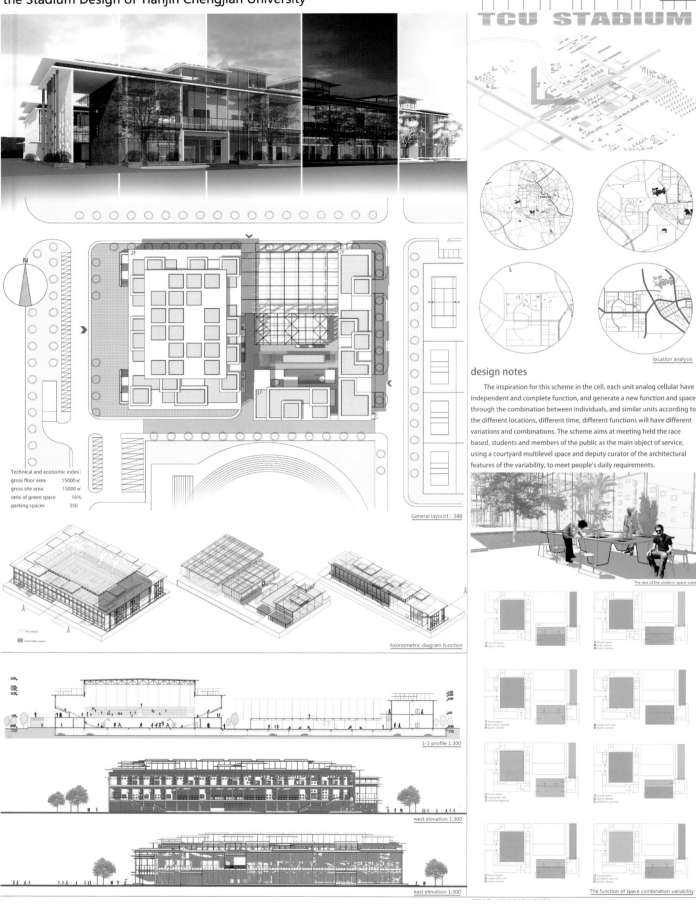

the Stadium Design of Tianjin Chengjian University

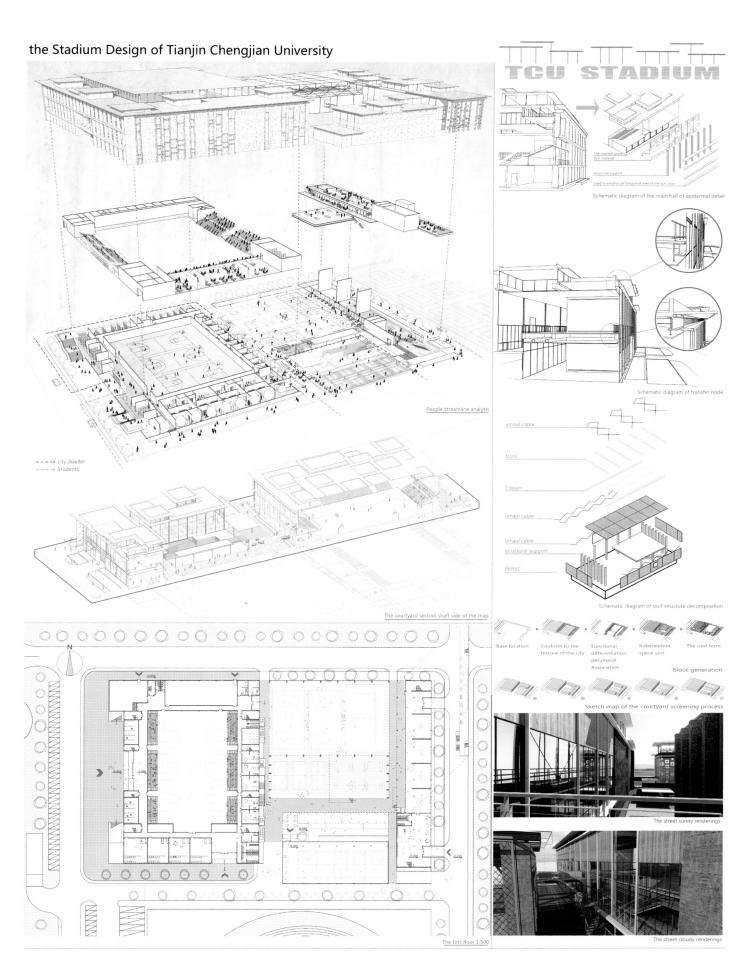

University: Tianjin Chengjian University
Designer: Zhang Jingru, Chen Keming, Peter Andersen, Janci Durza, Matds Rostrup
Tutor: Yan Li
Course name: the Stadium Design of Tianjin Chengjian University
Finished time: Jun., 2015
Exchange institute: VIA University College

Honorable Mentian

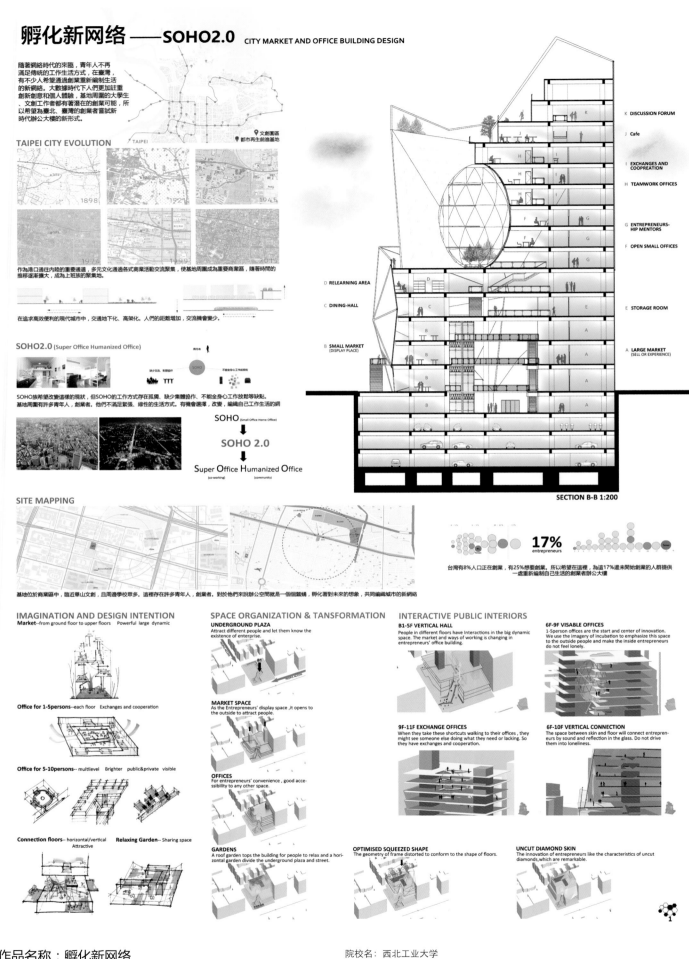

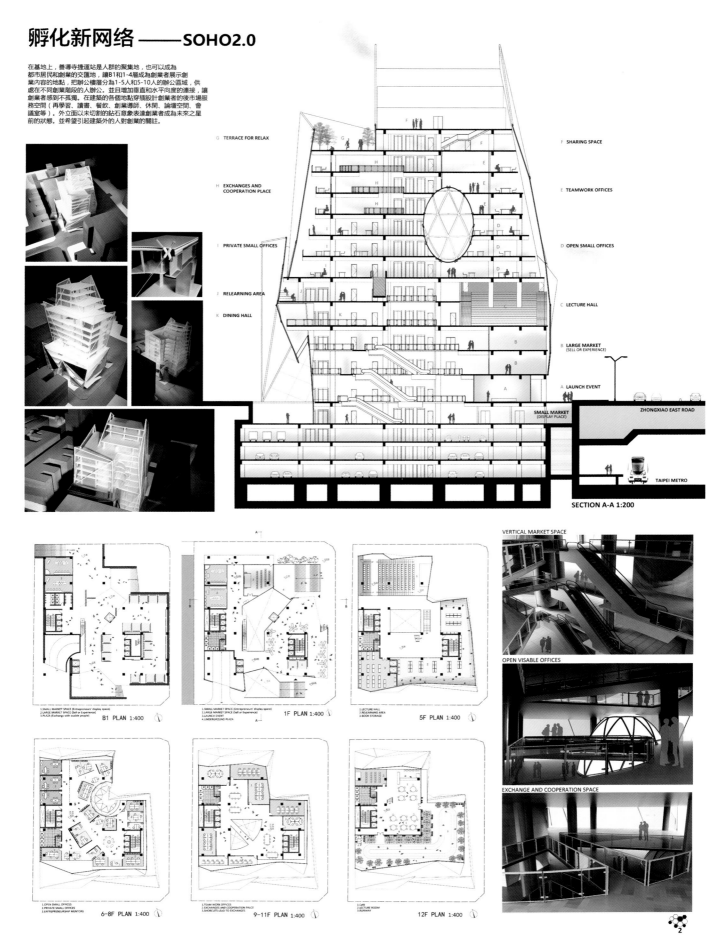

ACTIVATION
RURAL PUBLIC SPACES JOINT WORKSHOP

设计主题：以浙江省安吉县无蚊村作为设计基地，通过"微建筑"的介入，在强调小规模、地方性、易管理、生态型的基本理念下，拟改善乡村的公共设施、激活乡村的活力。

TOPIC: Design base is in Anji county, Zhejiang Province, Geng Village. The design through "micro-architecture" of the intervention, in emphasis on small-scale, local, easy management, basic ecological concepts, intended to improve the vitality of rural public facilities, activation countryside.

课程简介：要求同学们选定几个小地块，分别做一个校车巴士站、公厕、村民休息亭子，设计中需要使用一些地方材料，强调小规模、地方性、易管理、生态环保等。

COMMENTS: The design require the students to selected a few small plots. According to the theme to design a school bus station, a public toilets and a rest pavilion. The design require the students to use a number of local materials, emphasiz the small scale, local, easy management, ecological environment, etc..

课程总结：这次国际联合教学，我们进行了基于场地、材料、构造等方面的详细研讨与设计，对同学们拓展建筑视野、深入理解建筑内涵产生了积极的意义。

SUMMARY: In the international joint teaching, we carried out a detailed study and design based on space, material, structure and so on. Teaching has a positive significance for students to expand the vision of the building, in-depth understanding of the connotation of the building.

小建筑，微社区 01
——里庚村公共空间的激活与设计

COURSE	SURVEY	ANALYSIS	DISCUSSION	ITRIM	FINAL
2015.05.25 2015.05.29	2015.05.25	2015.05.26	2015.05.27	2015.05.28	2015.05.29
HANGZHOU	HANGZHOU	HANGZHOU	HANGZHOU	HANGZHOU	HANGZHOU

ANALYSIS

设计背景 BACKGROUND

里庚村位于安吉郵吴镇，杭州市的西北方向，村庄依山而建，周围群山环绕，民居和公共空间呈线性分布。

基地定位 / 地形地貌 / 村庄布局

村落景观 / 产业状况 / 茶园 / 稻田 / 手工业 / 休闲服务业 / "昌硕故里"文化

地方性材料 LOCAL MATERIALS

竹 Bamboo / 夯土 Rammed Earth / 砖 Brick / 瓦 Tile / 石 Stone

基地现状 PRESENT SITUATION

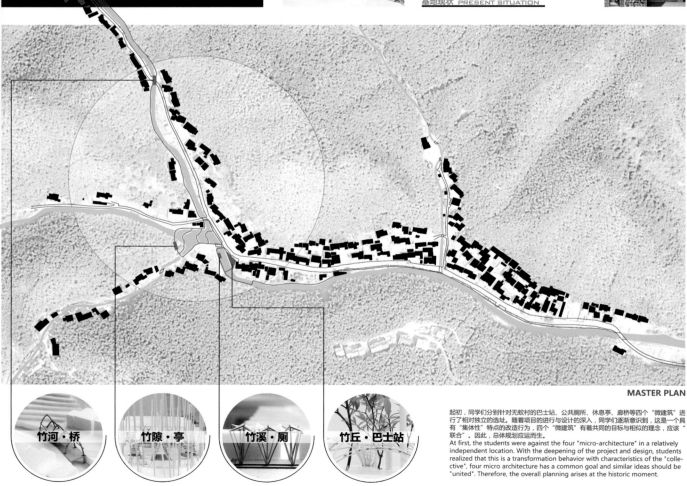

MASTER PLAN

起初，同学们分别针对无蚊村的巴士站、公共厕所、休息亭、廊桥等四个"微建筑"进行了相对独立的选址。随着项目的进行与设计的深入，同学们逐渐意识到，这是一个具有"集体性"特点的改造行为，四个"微建筑"有着共同的目标与相似的理念，应该"联合"。因此，总体规划应运而生。

At first, the students were against the four "micro-architecture" in a relatively independent location. With the deepening of the project and design, students realized that this is a transformation behavior with characteristics of the "collective", four micro architecture has a common goal and similar ideas should be "united". Therefore, the overall planning arises at the historic moment.

竹河·桥　竹隙·亭　竹溪·厕　竹丘·巴士站

作品名称：小建筑，微社区——浙江里庚村公共空间的激活与设计
Micro-architecture

入围奖

院校名：浙江大学
设计人：纪敏，闫嘉，沈昊，徐丹华，安秉君，吴杭冬，温茜玥，傅嘉言，孙姣姣，曾雨婷，戚骁锋
指导教师：孙炜玮，罗卿平，贺勇
课程名称：浙江大学—瑞士卢加诺南方应用技术大学联合workshop
作业完成日期：2015年05月
对外交流对象：瑞士卢加诺南方应用技术大学

小建筑，微社区
——里庚村公共空间的激活与设计

02

竹河·桥 | BRIDGE

GROUP 1

小组成员
沈昊、徐丹华、Stefano、Sophia

设计说明
在以"竹"业为生的无蚊村里，竹子常被直接放置于小河之上，同时作为供人通行的临时小桥。受此启发，我们期望在目前无蚊村的非活跃地段建造一座风雨竹桥，激活区域活力。在该地段，村庄主要的步行道通过一座混凝土板桥由东岸转至西岸，是重要的交通转折点。我们依托这个场地扩建，既能够满足了村民过河上山的生活需求，又能够提供给村民停留休憩的交往场所。

成员合影

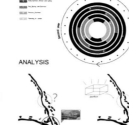

ANALYSIS | EXSITING BRIDGE | PLAN | STRUCTURE | SECTION

竹隙·亭 | PAVILLION

GROUP 2

小组成员
闫嘉、纪敏、Malpetti Alessandro

设计说明
我们利用两座步行桥和一条步行道连通了中心的公共空间。步行道穿过竹林，大大小小的竹亭沿着步行道两旁错落布置，保证了每个竹亭的良好视线。我们更像是搭建了一个小型公园，期望可以为村民、游客提供更有趣的体验——拾阶而上，漫步林中，择亭而憩，取景于心。坐在亭子中，透过竹林的缝隙，看着远山、近水和村庄里的行人；站在河对岸，欣赏竹林中若隐若现的亭子。亭内亭外的不同视角给许多有趣的故事提供了发生的可能性。

CONCEPTION | MASTER PLAN | PLAN | FACADE | SIZE | SECTIONING

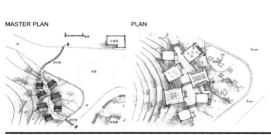
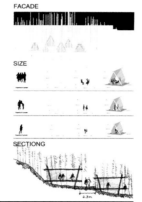

成员合影

竹溪·厕 | TOILET

GROUP 3

小组成员
吴杭冬、温茜玥、Stefao Mazzi、Nora

设计说明
我们的公共厕所"消隐于自然"。
我们希望公共厕所的竹结构能够既低碳经济，又方便施工。因此，我们根据厕所的体量，将竹子做成适宜尺度的三维立体杆件，通过对无蚊村村民和游客的人流量统计，最终集约成为两个男女混用的厕位。

成员合影

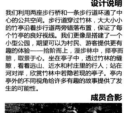

DESIGN SKETCH | STRUCTURE

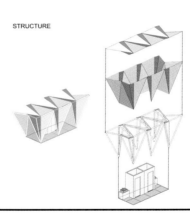
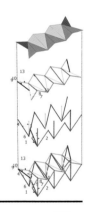

竹丘·巴士站 | STATION

GROUP 4

小组成员
安秉君、Flavio Facchini、Simona Scire

设计说明
巴士站的基地位于无蚊村的中心区域。该基地与完全开放的场地性质不同，内部巨大的乔木可以遮蔽阳光，但是无法遮挡雨水。巴士站旨在梳理场地条件的前提下，将三角形场地进行结构性覆盖，做一定的地景式抬升，划分出上下两个不同性质的空间，满足不同的使用需求。我们利用竹子弯折后"起拱抗压"的力学特征，构成了巴士站的主体结构。竹子绑扎数量为4根、2根、1根不等，将主体结构分成三个层级。

成员合影

MASTER PLAN | PLAN | SECTION | MODEL | DESIGN SKECTH

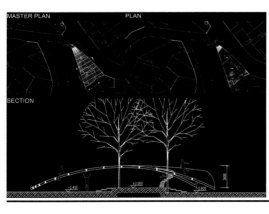

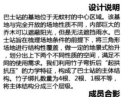

University: Zhejiang University
Designer: Ji Min, Yan Jia, Shen Hao, Xu Danhua, An Bingjun, Wu Hangdong, Wen Xiyue, Fu Jiayan, Sun Jiaojiao, Zeng Yuting, Qi Xiaofeng
Tutor: Sun Weiwei, Luo Qingping, He Yong
Course Name: Zhejiang University - University of Applied Sciences and Arts of Southern Switzerland Workshop
Finished Time: May, 2015
Exchange Institute: School of Architecture, University of Applied Sciences and Arts of Southern Switzerlandy

Honorable Mentian

WEAVING
Design of Spanish Craftwork Center

概念—编织 |
Concept—Weaving

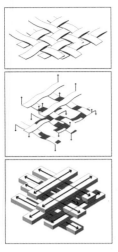

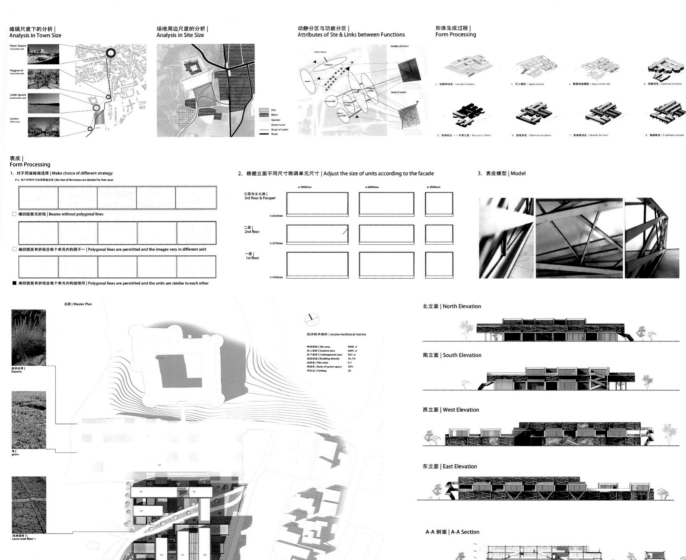

作品名称：西班牙传统手工艺博物馆设计
Design of Spanish Craftwork Center

院校名称：浙江大学
设计人：朱立涵
指导教师：金方，高裕江，吴璟
课程名称：浙江大学—San Pablo CEU 联合毕业设计课程教学
作业完成日期：2015 年 06 月
对外交流对象：西班牙 San Pablo CEU 大学

入围奖

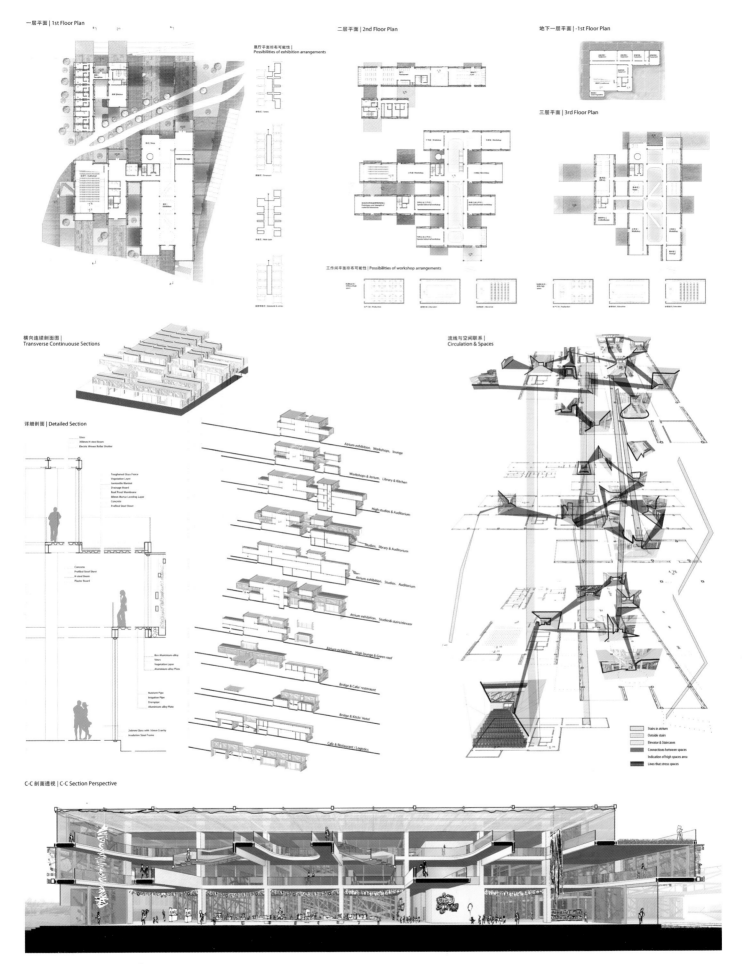

University: Zhejiang University
Designer: Zhu Lihan
Tutor: Jin Fang, Gao Yujiang, Wu Jing
Course Name: Zhejiang University - San Pablo CEU Joint Final Project Studio
Finished Time: Jun., 2015
Exchange Institute: School of Architecture, San Pablo CEU (Spain)

Honorable Mentian

New collective: Market, a give and take situation
A kind of derivative relationship

At the beginning of life, we come to this world in the purest and simplest way. Along with the endless time, we see away, feel away, grow away, get more and more mature, more and more intricate. Then we give birth to another new life, we deliver what we have known and experienced to them. Human always repeat this law generation after generation. The old one can give birth to the new one, and the new one can also be improved from the old one. Just like the situation like give and take.

photograph by Hiroshi Sugimoto

The relationship among three layers
From my orginal thinking I get my new cillective : talk about the relationship among layers. For example, what is the relationship among three layers:
        3 layers of roofs , 3 layers of walls, 3 layers of floors
I choose to contect each layer by some holes, I want to use these holes as a space, not only a window or a door, so from each holr, it leads to some new walls. Just like the situation in my memory, evety part grows from thr basic one, so i regard the middlemost cube as my core space, it is not only the most important space to experience, but also the main struture in my mew collective, The other tow cubes hang over this main struture layer by layer.

 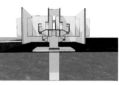
 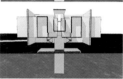

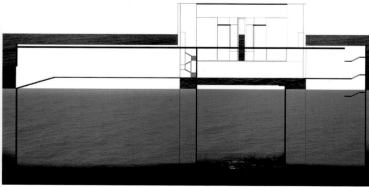

作品名称：美术馆
Art Museum

院校名：中央美术学院
设计人：毕拓
指导教师：Per Olaf Fjeld, Rolf Gerstlauer, Lisbeth Funck
课程名称：The new collective, Market, a give and take situation
作业完成日期：2015年06月
对外交流对象：挪威奥斯陆建筑与设计学院

入围奖

An art museum

It is an art museum made of concrete . Look afar,it looks like a rock on a peaceful sea.It contacts the land through a long bridge just like a invitation .Walk along this bridge,you won't know what you will see in this building.Get to the entrance,you can choose to walk continue or have a rest on the stairs ,here is open to the sea and sky,you will see and feel everything before your eyes. If you choose to walk continue you will experience a 30m long close corridor ,you can feel some light at the end of it,this lead you to enter to the building.Here, you will see the stairs at tow sides,climbing stairs you can see the seawater under your foot .Then you can get to the second layer,this exhibition layer is devided to tow parts, you can experience 2 close and 2 open exhibition spaces at each part,if you want to get the other part , you will have to climbing some stairs to get to the third layer first ,then get down to the other side.In the middle of each partf,there is a space for you to see the water again,here you can think about the exhibition or anything you think about at this moment.On the third layer, it is a public space for you,here you can walk along the main structure to see all the layers you have experienced ,at the end of this bridge, you can find a stairs to get down to the platform,as the wall is 3m high from the sea,so you can see all the scenery.Some time here will be a mini music concert.

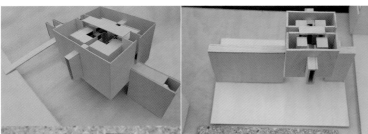

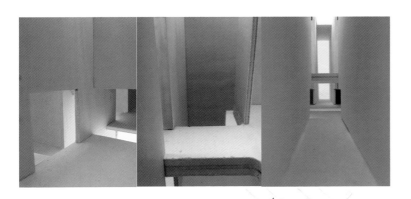

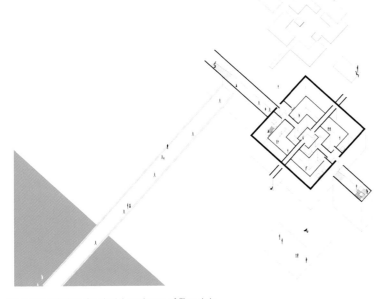

University: China Central Academy of Fine Arts
Designer: Bi Tuo
Tutor: Per Olaf Fjeld, Rolf Gerstlauer, Lisbeth Funck
Course Name: The new collective,Market,a give and take situation
Finished Time: Jun., 2015
Exchange Institute: The Oslo School of Architecture and Design

Honorable Mentian

New Collective
An open structure for the citizens and nature

01

The Construction of Dougong

Dougong is an important element in traditional Chinese architecture, and a reflection of the balancing wisdom integrating architecture and natural laws. I was inspired and hoped to achieve the constraint between structures and the balance among blocks.

The application of grey spaces in traditional Chinese architecture blurs the boundary between interior and exterior spaces with a softening touch and makes the transition from architecture to nature smoother.

Considering the characteristics mentioned above, I developed my prototype, in hope of creating a more flexible open space with adaptability.

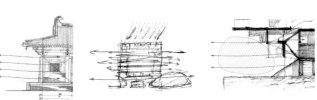

The purpose of the design is to research the relationship between nature and architecture. I was reflecting what kinds of architecture can be integrated into nature seamlessly, after which people in the situation can experience the influence nature has on architecture. In the end, I hope the space structure can be applied not only in a purely natural situation, but also in urban environment.

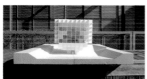

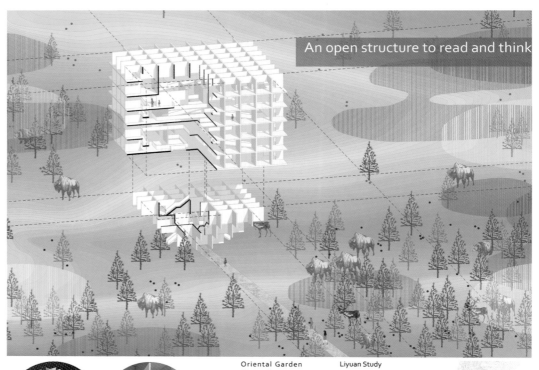

An open structure to read and think

Oriental Garden (Ryoan Ji)   Liyuan Study

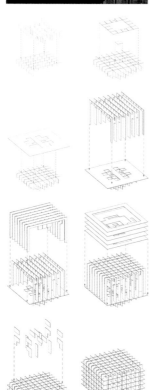
Building Process

The natural context of the design should be in the woods. After going through the dense woods and covering the distance of ups and downs, people reach this Shangri-La. The design is to be reincarnated into a treehouse made of wood, whose surface is intended to weathered with the marks of wind, rain, and snow. Here is the place where reading and thinking are done, and it is here that people forget if they are in an architecture or nature.

作品名称：新集体
The New Collective

院校名：中央美术学院
设计者：王子轩
指导教师：Per Olaf Fjeld, Rolf Gerstlauer
课程名称：The new collective, market, a give and take situation
作业完成日期：2015年05月
交流院校：挪威奥斯陆建筑与设计学院

入围奖

An open structure to be displayed and visited

02

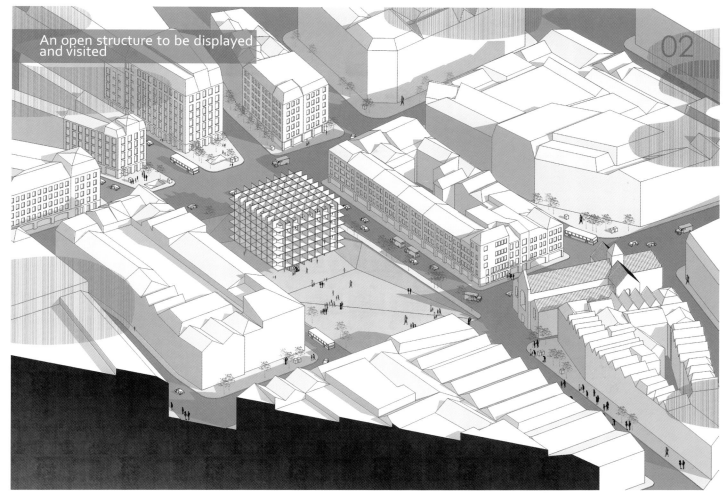

The site is located in city centre, which become a frame of street scenes.

The urban context of my design is the city centre of Brussel. I transformed my space design into a public space for sculpture display. Specific functions are achieved in the design while the harmony between the city and architecture is maintained.

The open space does not limit the viewing experience to the architecture and exhibits. Instead, a platform for a better view of the city is thus created, and the fun is that people, the city, and the exhibits are in an enclosed circle of viewing and being viewed. Meanwhile, the compartmentalization ensures privacy to a certain level. Thus, the integration with city environment and culture is achieved.

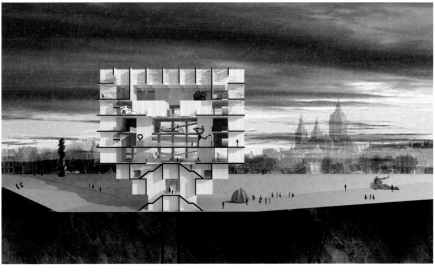

University: China Central Academy of Fine Arts
Designer: Wang Zixuan
Tutor: Per Olaf Fjeld, Rolf Gerstlauer
Course Name: The new collective, market, a give and take sitnation
Finished Time: May, 2015
Exchange Institute: The Oslo School of Architecture and Design

Honorable Mentian

WohnKulturen Vergleich zwischen traditionelle und moderne Wohnhaus

Entwickelt, um internationale Einwohner zu gewinnen, ist Park 5 ein High-End- Entwicklung nach der kosmopolitischen Ballungs Gemeinden in Städten wie Paris und New York gestaltet. Das Hotel liegt in der Nähe von Chao Yang Park in Peking, schafft dieses Wohn- und Geschäftsentwicklungseinem exklusiven Umfeld.

Der Rahmen des Gebäudes schafft ein Gefühl von Exklusivität ansprechbar, dynamische Komplexität und luxuriösen Raffinesse. Perfekte Details machen das Design Appell an alle Stadt Eliten und bietet die entspannte, hochwertige persönlichen Lebensraum.

Ein Architekt baut Hauptstadt, wie ein ebenes Quadrat mit einer Seitenlänge von neun Li, jede Seite hat drei Tore. Die Stadt verfügt über neun vertikalen und horizontalen neun großen Straßen Breite der Straße sind alle gleichzeitig mit neun Wagen. Die linke Seite des Palastes (Ost) ist die Jongmyo und die rechte (West) ist ein Sajik. Vor dem Palast ist der Minister Gottesdienst, gefolgt von dem Markt.
《The Ritual Works of Zhou·Kao Gong Ji》

Peking ist die Hauptstadt der China. Die Bevölkerung ab 2013 war 20.150.000. Die Stadt selbst ist die 3. größte in der Welt und in Nordchina befindet.
Yuan Ming und Qing-Dynastie Innenhof reift, der Yuan- Kaiser Land in die Hauptstadt der wohlhabenden, Beamte zu gehen, so dass sie verwendet werden, um Häuser zu bauen. Das Bild zeigt eine Draufsicht auf den Hof der Qing Adelsresidenz.

- Back Yard
- West-side Room
- Courtyard
- Chaoshou Corridor
- Back Room
- Front Yard
- Back Cover Room
- Main Room
- East-side Room
- Festoon Gate
- Screen Wall
- Gate

 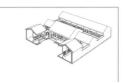

Back Cover Room — Bedroom for servants and unmarried daughters
Side Room of Main Room — Study Room for Master
Main Room — Living Room and Bedroom for Master
West-side Room — Living Room and Bedroom for Second son and his family
East-side Room — Living Room and Bedroom for Eldest son and his family
Back Room — Kitchen Dining Room Meeting and others

Jedes Zimmer hat 1,3,5,7 jianzwischen unterteilt.

A- Innen (privat)
B- halboffenen (halb-öffentlichen)
C- öffnen (öffentlich)

Der kleine Hof verfügt über 13 Häuser, und die größere Mai hat 30 oder mehr. wing-Raumes Rückwand ist die Wand der Hof, und dann bauen die einen Mauern um die Ecke, einem großen Innenhof mit einer Mauer von der Außenseite umgeben, hohen Mauern ohne Fenster, das zeigt eine defensive. Die ganze Familie zusammen zu bekommen in den Innenhof mit bequemen. Wenn Sie Tür in der Nacht geschlossen, ist es sehr ruhig und eignet sich für Familien zusammen leben.

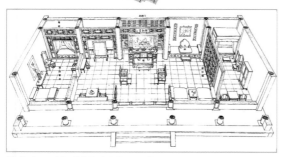

Altes Haus Layout relativ unflexibel ist, in der Regel jedes Zimmer Layout folgt dem Prinzip der inter- Basis. Alte Menschen großen Wert auf Symmetrie und deshalb sind sie das wichtigste, wie die Mitte des Raumes, in der Regel für das Leben oder Gottesdienste genutzt. Und basierend auf der Ausrichtung unterschiedlich ist, sich auf beiden Seiten des Raumes für die Untersuchung verwendet, die von der Gastronomie oder Schlafzimmer.

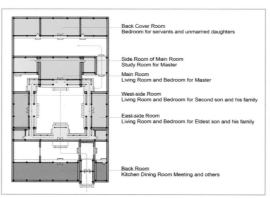

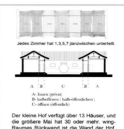

Dynastie	YUAN 1271-1368	MING 1368-1644	QING 1644-1912	Republic of China 1912-1949	R.P.China 1949-
Bereich	5400m²	weniger als 5400m²	größer/kleiner	einige sogar kleiner	einige sogar kleiner
			1 Familie	2-3 Familie 1 Familie	10-20 sogar mehr

Beijing Hutong meistens von Osten nach Westen, also meist von Norden nach Süden, eine Eigenschaft, die in der Innenstadt von Peking besonders deutlich ist Hof. Courtyard Tür ist in der Regel im Südosten geöffnet und nicht gegenüber dem Haupthaus.
Wenn Siheyuan liegt im Süden der Hutong befindet, und der Süden des Hofes ist nicht auf der Straße, die nicht geöffnet werden Hof bedeutet geöffnet, so dass die Tür an der Nordwest.
Zeit vergeht, wird die Popularität schnell aufwachsen, so dass die Art und Weise des Lebens in siheyuan hat sich geändert.

设计题目：中国住宅设计文化的传承
Wohnkulturen

院校名：中央美术学院
设计者：宋羽
指导教师：Prof.Dr.Christine Hannemann, Dr.Gerd Kuhn
课程名称：居住文化
作业完成日期：2015年08月
对外交流对象：德国斯图加特大学

入围奖

WohnKulturen Vergleich zwischen traditionelle und moderne Wohnhaus

Entwickelt, um internationale Einwohner zu gewinnen, ist Park 5 ein High-End-Entwicklung nach der kosmopolitischen Ballungs Gemeinden in Städten wie Paris und New York gestaltet. Das Hotel liegt in der Nähe von Chao Yang Park in Peking, schafft dieses Wohn- und Geschäftsentwicklungseinem exklusiven Umfeld.

Der Rahmen des Gebäudes schafft ein Gefühl von Exklusivität ansprechbar, dynamische Komplexität und luxuriösen Raffinesse. Perfekte Details machen das Design Appell an alle Stadt Eliten und bietet die entspannte, hochwertige persönlichen Lebensraum.

Location: Beijing (Chaoyang District)
Size: 38,000 sqm (73,000/96,000)
Scope of service: Schematic plan, architectural and landscape design
Project type: Residential & Commercial
Client: Vanke Corporation
Design Team: GBBN Beijing, etc.
Status: Completed 2011

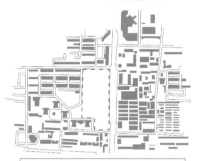

Vanke V Park Bezirk ist vor allem von ausländischen Handels-, Wohn- und Jule Ministerien zusammen. Relativ geschlossenen öffentlichen Privatsphäre auf die Bildung der drei Haupt Innenhof. Geländer wie zeitgenössische chinesische Gebäude in einem Merkmal, das angewendet wird.
Als Chinas Hauptstadt und zieht eine große Zahl von jungen Menschen kommen hierher, und entwickeln ein Geschäft. Anzeigen angespannt Stadt hat sich zu einem großen Problem, so wählen Sie die Form der Zusammenarbeit Vermietung von modernen China als meine Hauptaufgabe der Analyse von Wohngebäuden.

- the fees of daily necessities, including water, gas and electricity
  no certainly rules
- share with others - living room - small - staff
- kitchen is very important in China
- share with others - kitchen - 3 or 4 types of the same kitchenware
  the frequency of use is sharply increased on weekends you may not have chance to use it in a proper time
- private space is not private - eg. sound \ close door
- unstable states - landlord - distempered ness of related law
- heating system - centrally heated (government) & floor heating system

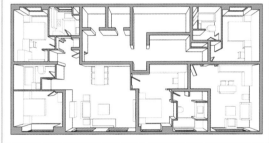

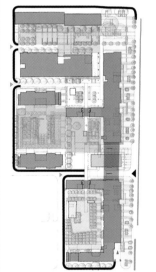

Modernes Gehäuse ist ganz anders, da die Co-Mieter der Grund, aus dem Eingang zum Wohnzimmer haben die Rolle des Service-Raum, Küche, der am häufigsten verwendeten öffentlichen Raum, gefolgt von dem Restaurant, dem unteren Wohnzimmer angenommen.
Co-Zimmer-Vermietung in Funktionen sind weitere Definition gegeben, da Haushalte sind nicht vertraut mit einander, relativ gesehen wird wählen Schließen der Tür die meiste Zeit allein im Schlafzimmer, das Schlafzimmer und daher auch als Lebenssinnzu handeln Rolle.

Dies ist ein 180 m2 Familienhaus, aus dem Wohnzimmer ins Schlafzimmer schrittweise Stärkung der Privatsphäre. Unter normalen Umständen bieten 3-4 für die Mieter zu bedienen, einschließlich der Master-Schlafsphäre mit Bad südöstlicher Richtung, und die verbleibenden zwei Schlafzimmer teilen ein Bad. Häufigkeit der Nutzung von Wohnraum wird durch die Wechselwirkung zwischen reduziert die Mieter nicht vertraut sind, so dass manchmal der Vermieter eine vierte Schlafzimmer im Wohnzimmer auf ihre eigenen zu bauen, in der Regel besetzen die Hälfte des Wohnzimmers.

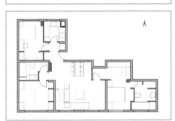

Aus historischen Gründen ist China nicht mehr in der modernen Häuser der Anbetung und Stühle, die Religionsfreiheit vorhanden, sondern macht Haushalt für die relevanten Aspekte der Raumaufteilung variieren.
Allerdings ist die Position und die Orientierung des Spiegels Anzeige, Mieter in der Regel mehr Aufmerksamkeit. Spiegeln Symbolik in den fünf Elementen der besonderen und daher besondere Aufmerksamkeit widmen sollten.

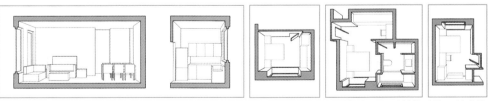

Für Co-Vermietung, die Küche hat sich zu einem wichtigen Kern ersetzen das Wohnzimmer Zimmer. Da Teilung zwischen Benutzer sind nicht miteinander vertraut und wird daher nicht langer Aufenthalt im Wohnzimmer.

University: China Central Academy of Fine Arts
Designer: Song Yu
Tutor: Prof. Dr. Christine Hannemann, Dr. Gerd Kuhn
Course Name: Wohnkulturen
Finished Time: Aug., 2015
Exchange Institute: Universität Stuttgart

Honorable Mentian

城市设计

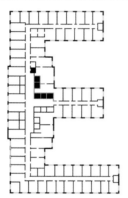  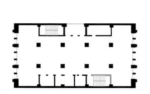 

作品名称：城市设计
Urban Design

院校名：中央美术学院
设计人：闫玉琢
指导教师：Cintya Eva Sanchez Morales
课程名称：设计工作室 3
作业完成日期：2015 年 07 月
对外交流对象：奥地利因斯布鲁克大学建筑学院

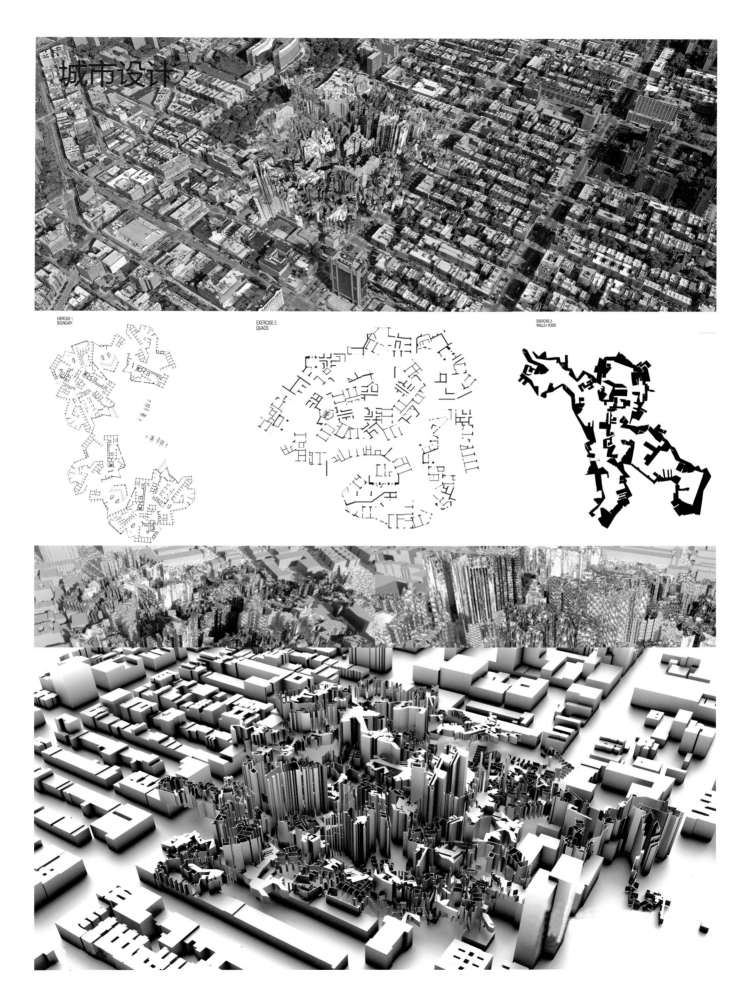

University: China Central Academy of Fine Arts
Designer: Yan Yuzhuo
Tutor: Cintya Eva Sanchez Morales
Course Name: Workshop 3
Finished Time: Jul., 2015
Exchange Institute: University of Innsbruck

Honorable Mentian

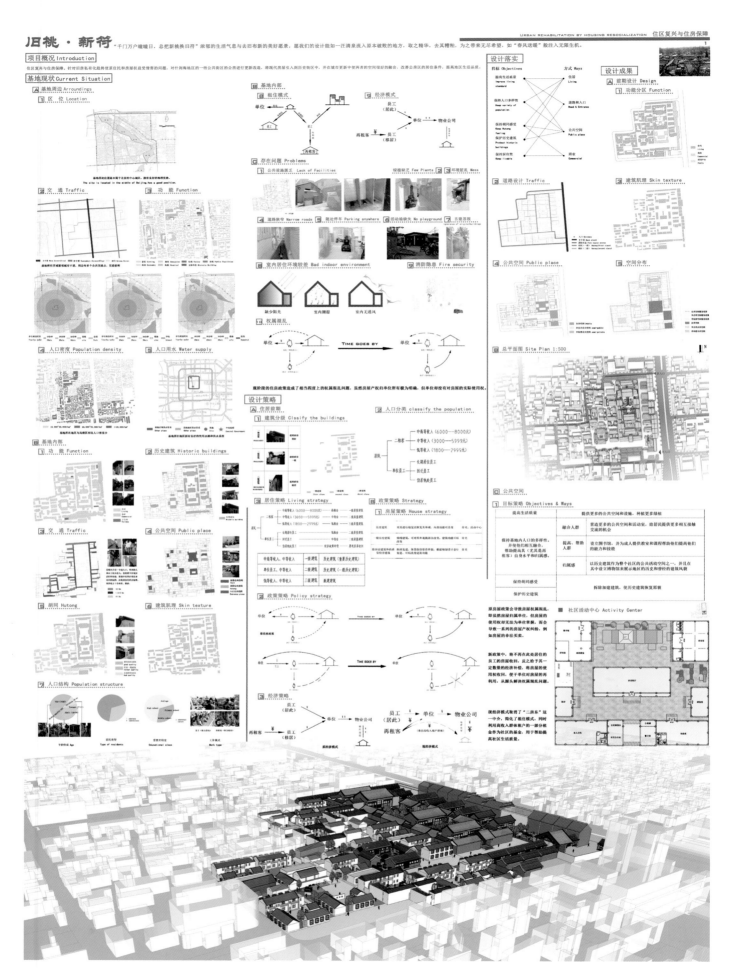

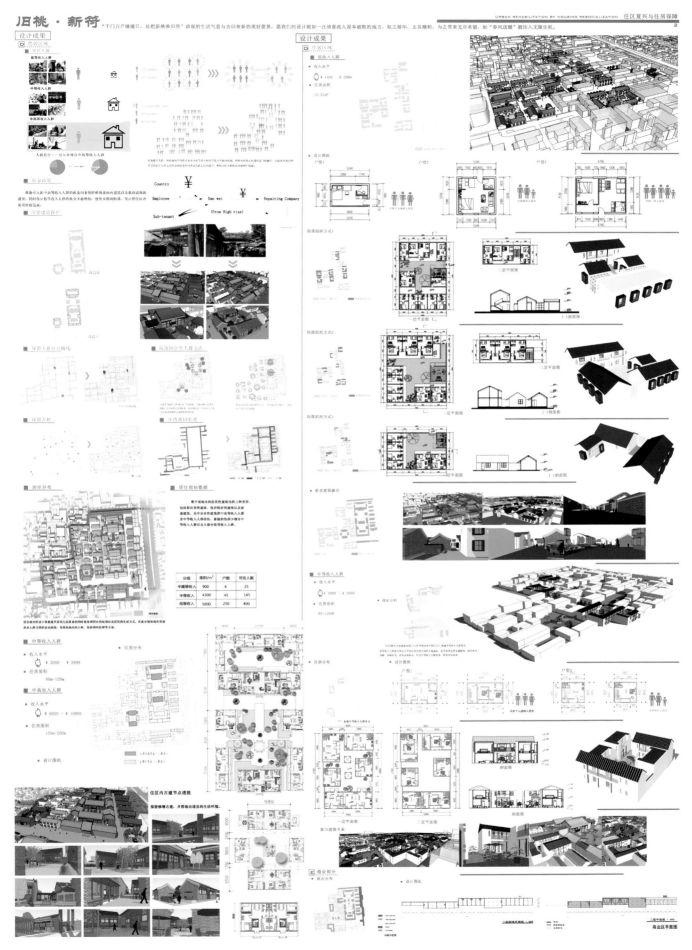

University: Beijing University of Technology
Designer: Zhou Xinyi, Li Chenxi, Wang Lijing, Zhu Baoyu, Li Meng, Zhou Yecheng, Li Ding, Yang Fujun
Tutor: Hui Xiaoxi, Maurice, Harteveld, Jiang Bing
Course Name: BJUT-TU Delft Joint Urban Design Studio
Finishied Time: Jul., 2015
Exchange Institute: Faculty of Architecure and the Built Enviornment, Delft University of Technology

Honorable Mentian

LAYERING Developing Federation Square East, Melbourne

Master Plan 01

See you there to embodied experience

In the age of private digital telecommunications so pervasive and intensive around us, 24/7, all time and everywhere, what is the meaning of a public space in the city today?

Melbourne is regarded as a city of experience, in order to find the answer, my partner and I use our body to see, to smell, to hear, to feel, to taste and to explore Melbourne, in and through the laneways, the squares, the arcades, the cafés, the park and the riverfront; and we find that the Australian people also fascinate in doing out-door activities by themselves, like going out to meet face on face, enjoying the sunshine and the views, participating a 'civic' excitement, running around the city, feeding the birds, to name just a few. Therefore, 'embodied experience' might be one simply answer to the social problem which cannot be replaced by digital telecommunications.

Then we record our observation of various types of people's daily activities. We just want to create some interesting spaces which could combine these different activities and events to help people embodied experience outside. After that we discover different level of layers, the floors-they might be overlapped or separated-can help us to realize our goals. As a result, we have our concept-layering.

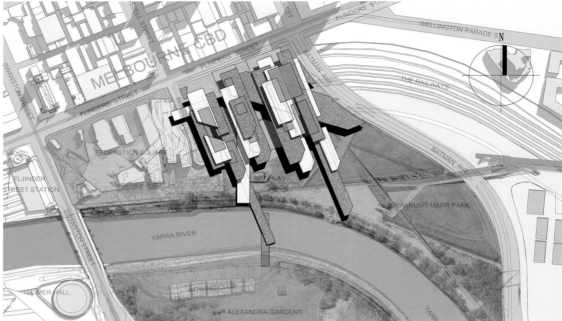

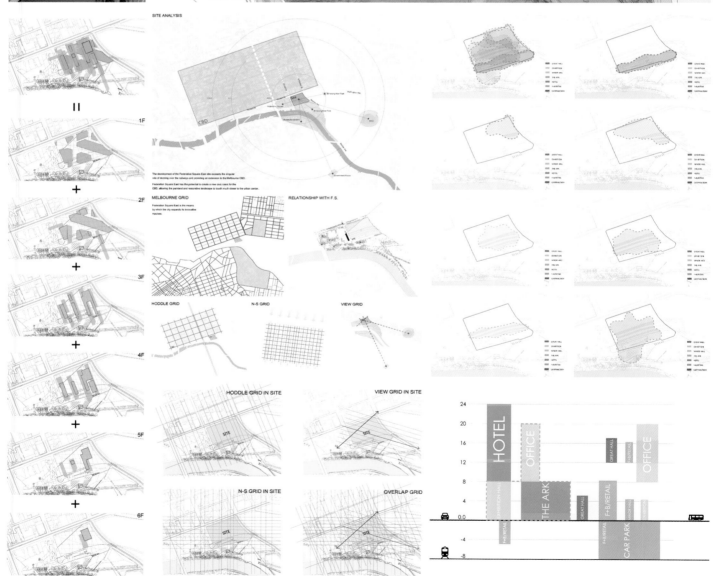

作品名称：墨尔本联邦广场东侧地块城市设计（1）
Developing Federation Square East, Melbourne（1）

入围奖

院校名：南京大学
设计人：胡任元，吴嘉鑫
指导教师：华晓宁，朱剑飞，D.Bates, J.Turnbull
课程名称：南京大学—墨尔本大学强化工作营
作业完成日期：2015年07月
对外交流对象：澳大利亚墨尔本大学建筑、房屋和规划学院

LAYERING Developing Federation Square East, Melbourne

Urban Design 02

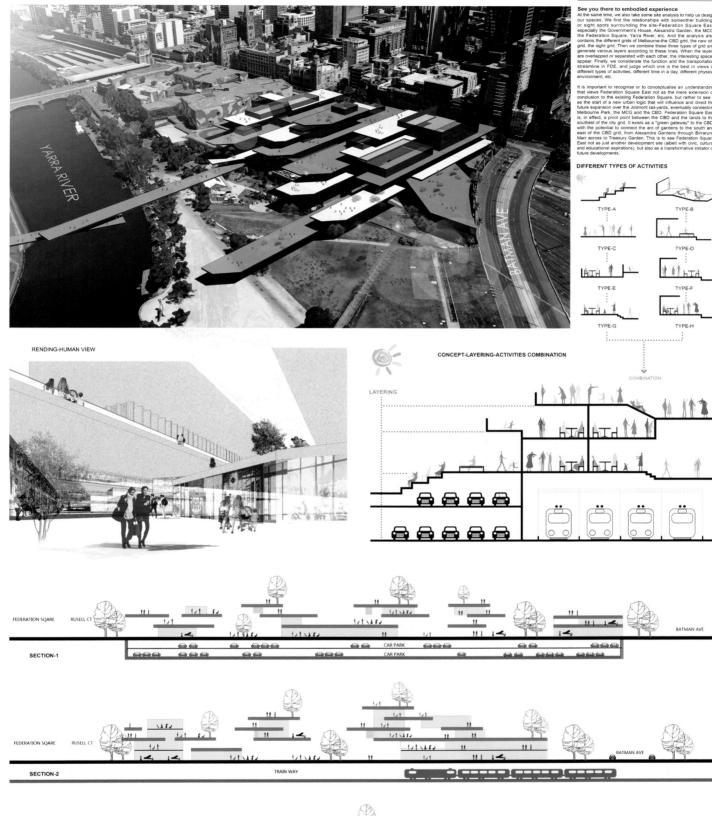

See you there to embodied experience

At the same time, we also take some site analysis to help us design our spaces. We find the relationships with someother buildings or sight spots surrounding the site-Federation Square East, especially the Government's House, Alexandra Garden, the MCG, the Federation Square, Yarra River, etc. And the analysis also contains the different grids of Melbourne-the CBD grid, the new city grid, the sight grid. Then we combine these three types of grid and generate various layers according to these lines. When the layers are overlapped or separated with each other, the interesting spaces appear. Finally, we considerate the function and the transportation streamline in FDE, and judge which one is the best in views of different types of activities, different time in a day, different physical environment, etc.

It is important to recognise or to conceptualise an understanding that views Federation Square East not as the mere extension or conclusion to the existing Federation Square, but rather to see it as the start of a new urban logic that will influence and direct the future expansion over the Jolimont rail-yards, eventually connecting Melbourne Park, the MCG and the CBD. Federation Square East is, in effect, a pivot point between the CBD and the lands to the southeast of the city grid. It exists as a "green gateway" to the CBD, with the potential to connect the arc of gardens to the south and east of the CBD grid, from Alexandra Gardens through Birrarung Marr across to Treasury Garden. This is to see Federation Square East not as just another development site (albeit with civic, cultural and educational aspirations), but also as a transformative initiator of future developments.

University: Nanjing University
Designer: Hu Renyuan, Wu Jianxin
Tutor: Hua Xiaoning, Zhu Jianfei, Donald Bates, J.turnbull
Course Name: Nanjing - Melbourne University Design Studio
Finishied Time: Jul., 2015
Exchange Institute: University of Melbourne

Honorable Mentian

SEE YOU THERE
Developing Federation Sqaure East, Melbourne 2015

01

As the digital telecommunication becoming more and more universal, our daily life is full of cellphones, computers and other electronic devices. The devices like telephone, cellphones in one way can make our relationship with friends or colleague more intimate, it also in another way put our relation in an untouchable way (we may just only talk by phone). How do we to face these question, How do we handle these issue.

Melbourne is a city of experience, The art and culture, pubs, food and wine, spectator sports are amazing and fantastic in the city, especially laneways which have a variety of use, such as cafe, breakfast, pubs and other interesting place. When you walking through a main or second street in the CBD in Melbourne, you will discover such hidden space at night. The lively atmosphere, the waiter with passion will inspire your enthusiasm to have a cup of a coffee or a bottle of wine. You may have chat with a strangers or meet with your best friend, all you have to do is take easy and enjoy the moment at this time.

Laneways

In Melbourne, each block have there own laneways as the time past by. As the picture shown, the line in orange color is laneways. These line become part of city texture.

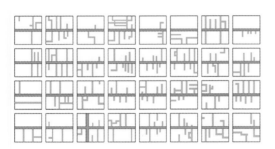

City railway break the texture from city centre to yarra river. There are exsisting project like Federation Sqaure and our designing space—Federation Sqaure East. However, the railway extends out our site, as the left picture show, there will be more project be built upon the railway. How to figure out a solution is important.
Metabolism inspire us and we want to build a megastructure upon the railway to continue the city texture.

The development of the Federation Square East is our site for designing, it allows for the expansion and provision of cultural and civic amenities, supplementing those established by Federation Square, while offering the opportunity for new types of public-private facilities. We want to have a continues way to figure out how to build in the railway which have broken the city fabric.

Site Introduction and the main streets of the city centre

Added the Laneways

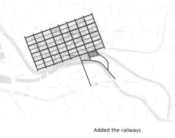

Added the railways

Added the green space and clutre precinct

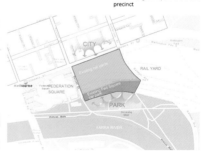

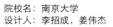

作品名称：墨尔本联邦广场东侧地块城市设计（2）
Developing Federation Square East, Melbourne (2)

入围奖

院校名：南京大学
设计人：李招成，姜伟杰
指导教师：华晓宁，朱剑飞，D.bates, J.Turnbull
课程名称：南京大学—墨尔本大学联合设计课程教学
作业完成日期：2015年07月
对外交流对象：澳大利亚墨尔本大学建筑、房屋和规划学院

SEE YOU THERE
Developing Federation Sqaure East, Melbourne 2015

02

Process Analysis

Exsiting railways

Laneways from city centre

The base buiding area

Revise of base building area

Introduce the outside walkways

Detail of the base structure

University: Nanjing University
Designer: Li Zhaocheng, Jiang weijie
Tutor: Hua xiaoning, Zhu Jianfei, D.bates, J.Turnbull
Course Name: Nanjing-Melbourne University Design Studio
Finished Time: Jul., 2015
Exchange Institute: University of Melbourne

Honorable Mentian

CHAOS IN URBAN DESIGN: MAPPING x HAMALS 01

City scale: Mapping the logistics industry in Hangkou

Since the trading port opened in 1860s, Hankou has turned into one of the most important logistics sites of central China. Hanzheng Street Area, close to the junction of Yantze river and Han river, became the most prosperious area, and there formed kinds of dock culture and market culture. Hamals are one indispensible part of this culture.

Recently, with the reconstruction of Hankou old town being proposed, Hanzheng Street Area has been considered as commodities exchange center. In the planning, it is easy to notice that the insentive markets, as new financial area, will take place of old commercial streets. Therefore, the life and work of hamals will also transform to follow urbanization process--- from long-distance transport in narrow lanes to short-distance transport in markets.

Hanzheng Streets Area: The center of logistics in Wuhan

Hamals: wokers who carry goods bewteen old town and markets or logistics site.

Community scale: Mapping place for work and life of hamals

In the investigation, we found that hamals **tend to assmeble in the boundary space between market(workplace) and old town(home)**, where they can not only wait for work but also participate in the city life.

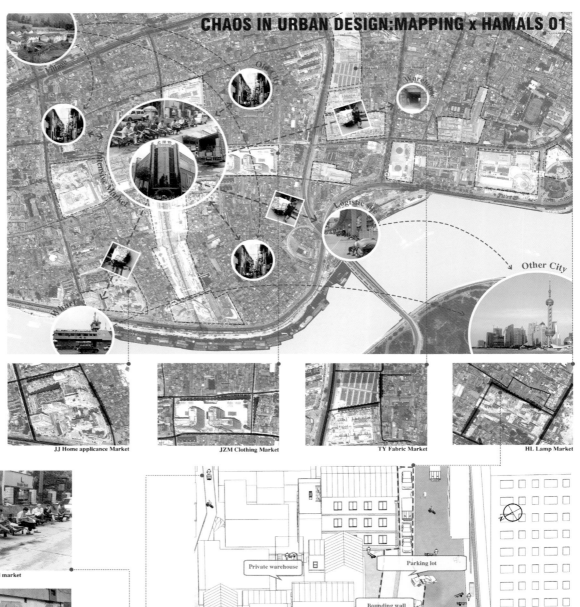

Place for assemmbing: Boundary of old town and market

Place for working: Market warehouse or logistics site

Place for carrying: Lanes in old town

Place for city life: Old town

作品名称：探寻花楼街——板车组
Investigating Hualou Street

入围奖

院校名：武汉大学
设计人：周韦博，康雅迪，张林凝，李晶，王青子，青妍，王子钦
指导教师：何志森，胡小青，张点
课程名称："规划的混乱"Mapping 工作坊
作业完成日期：2015年07月
对外交流对象：澳大利亚墨尔本皇家理工大学，非正规工作室（澳大利亚）

Human scale: Life and work of hamals

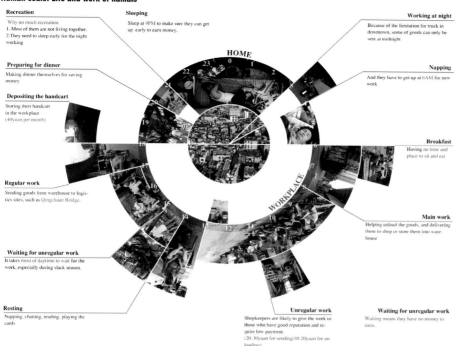

Recreation — Why no much recreation 1. Most of them are not living together. 2.They need to sleep early for the night working

Sleeping — Sleep at 9PM to make sure they can get up early to earn money.

Working at night — Because of the limitation for truck in downtown, some of goods can only be sent at midnight

Preparing for dinner — Making dinner themselves for saving money

Napping — And they have to get up at 6AM for new work

Depositing the handcart — Storing their handcart in the workplace (40yuan per month)

Regular work — Sending goods from warehouse to logistics sites, such as Qingchuan Bridge.

Breakfast — Having no time and place to sit and eat

Main work — Helping unload the goods, and delivering them to shop or store them into warehouse

Waiting for unregular work — It takes most of daytime to wait for the work, especially during slack season.

Resting — Napping, chatting, reading, playing the cards

Unregular work — Shopkeepers are likely to give the work to those who have good reputation and require low payment. (20-30yuan for sending/10-20yuan for unloading)

Waiting for unregular work — Waiting means they have no money to earn.

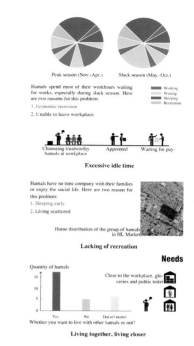

Peak season (Nov.-Apr.) Slack season (May.-Oct.)

Hamals spend most of their workhours waiting for works, especially during slack season. Here are two reasons for this problem:
1. Economic recession
2. Unable to leave workplace

Choosing trustworthy hamals at workplace | Appointed | Waiting for payment

Excessive idle time

Hamals have no time company with their families or enjoy the social life. Here are two reason for this problem:
1. Sleeping early
2. Living scattered

Home distribution of the group of hamals in HL Market

Lacking of recreation

Needs

Close to the workplace, groceries and public toilet

Whether you want to live with other hamals or not?

Living together, living closer

Human scale: Strategies in carrying goods

Strategy to stretch ropes — Hamals used to cram one small box into well-piled goods to make sure that the ropes are tight enough.

Strategy to keep balance — Hamals used to pile up more goods on the back of handcart, so that human-handcart system could be a stable rectangle.

Strategy to locate box — Hamals used to pile up box with quarter bonds so that goods will not fall apart.

Strategy to locate fabric — Hamals tend to use rope to bind rolls of fabric layer upon layer.

Human scale: Reconstruction of handcarts

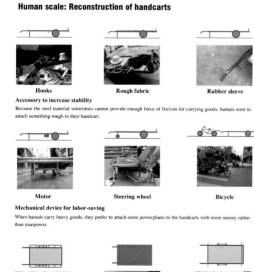

Hooks | Rough fabric | Rubber sleeve

Accessory to increase stability
Because the steel material sometimes cannot provide enough force of friction for carrying goods, hamals tend to attach something rough to their handcart.

Motor | Steering wheel | Bicycle

Mechanical device for labor-saving
When hamals carry heavy goods, they prefer to attach some powerplants to the handcarts with more money rather than manpower.

Steel bars | Plank | Bamboo cane

Increasing width to add loading capacity
In general, more loading capacity means more money in one carrying period. Therefore, hamals tend to increasing the width of handcarts to gain more loading capacity.

CHAOS IN URBAN DESIGN: MAPPING x HAMALS 02

University: Wuhan University
Designer: Zhou Weibo, Kang Yadi, Zhang Linning, Li Jing, Wang Qingzi, Qing Yan, Wang Ziqin
Tutor: Jason Ho, Hu Xiaoqing, Zhang Dian
Course Name: "Planned Chaos" Mapping Studio
Finished Time: Jul., 2015
Exchange Institute: RMIT Univeristy, Urban Informality Lab(AU)

Honorable Mentian

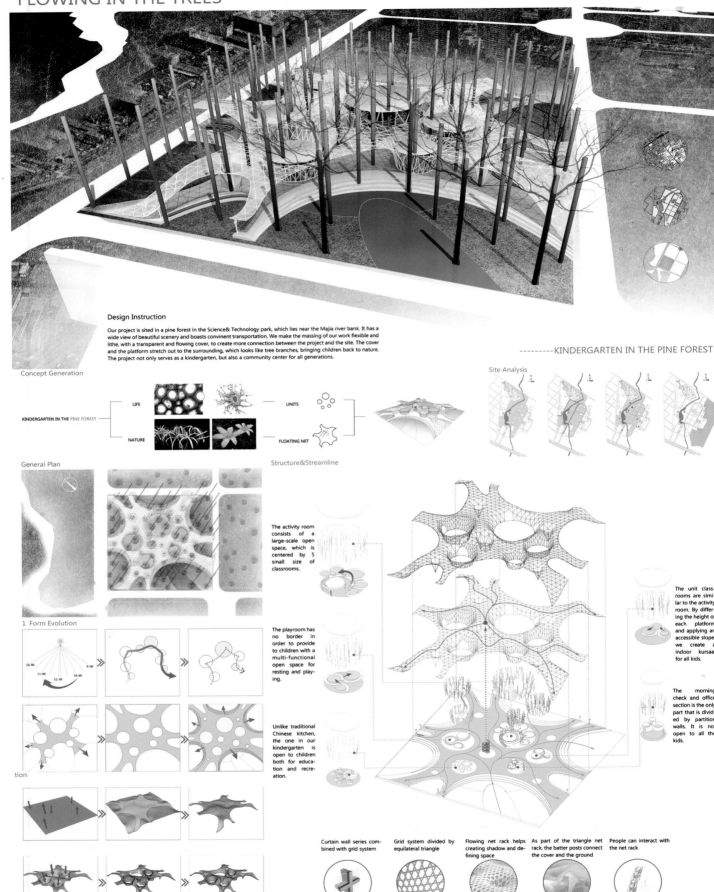

FLOWING IN THE TREES

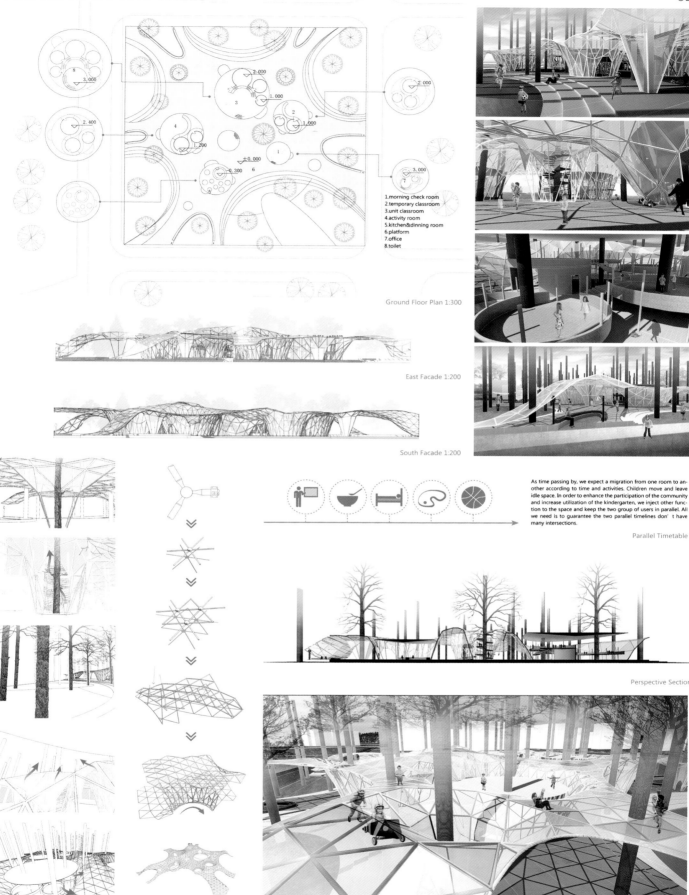

University: Harbin Institute of Technology
Designer: Sun Jiaqi, Du Jing, Zhao Weinan
Tutor: Bu Chong, Marta Barrera Altemir
Course Name: Harbin Institute of Technology. Summer School 2015
Finished Time: Jul., 2015
Exchange Institute: BAUM Architects

Honorable Mentian

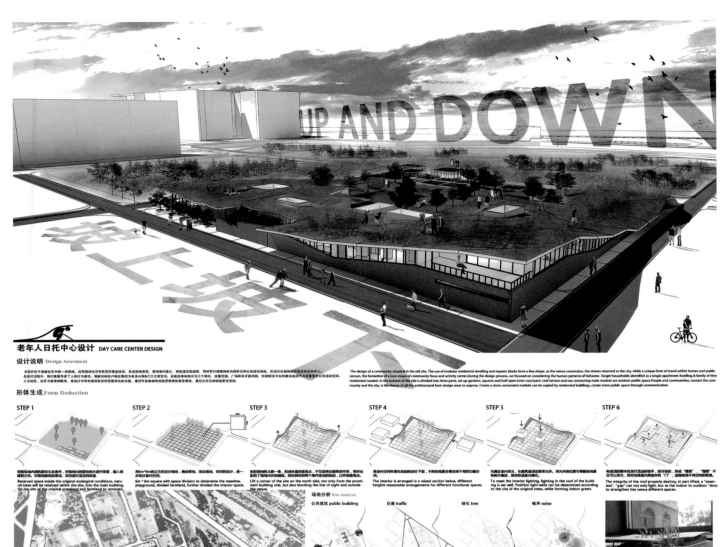

老年人日托中心设计 DAY CARE CENTER DESIGN

作品名称：老年人日托中心设计
Day Care Center Design

入围奖

院校名：哈尔滨工业大学
设计人：陆柏屹，王冠群，蔡恒屹，鲍润生
指导教师：吴健梅，史立刚，Mr. Miguel Gentil
课程名称：哈尔滨工业大学—西班牙BAUM建筑事务所联合设计课程教学
作业完成日期：2015年07月
对外交流对象：西班牙BAUM建筑事务所

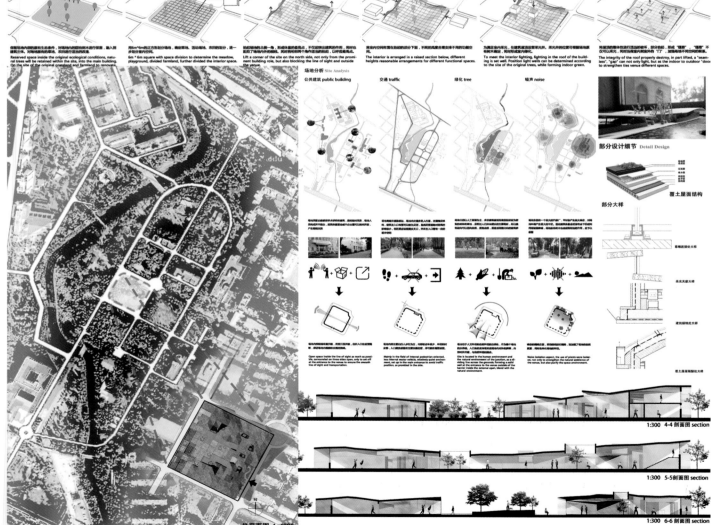

University: Harbin Institute of Technology
Designer: Lu Baiyi, Wang Guanqun, Cai Hengyi, Bao Runsheng
Tutor: Wu Jianmei, Shi Ligang, Mr. Miguel Gentil
Course Name: Harbin Institute of Technology - BAUM Architects Joint Design Studio
Finished Time: Jul., 2015
Exchange Institute: BAUM Architects

Honorable Mentian

2016 年中国建筑院校境外交流优秀作业名单

一等奖

序号	院校及院系全名	对外交流对象（国家、地区、学校及院系）	设计题目	课程名称	学生姓名	第一完成学生所在年级	指导教师姓名
1	哈尔滨工业大学建筑学院	中国文化大学	猴硐综合体——基于产业活化的选煤厂改造	开放设计	张相禹、魏娜、王艺凝	本科四年级	周立军、殷青、韩衍军
2	南京大学建筑与城市规划学院	英国剑桥大学建筑系	学术景观——剑桥水边仓库图书馆改造	南京大学—剑桥大学联合毕业设计	席弘	本科四年级	窦平平、Ingrid Schröder

二等奖

序号	院校及院系全名	对外交流对象（国家、地区、学校及院系）	设计题目	课程名称	学生姓名	第一完成学生所在年级	指导教师姓名
1	清华大学建筑学院	美国耶鲁大学建筑学院	天津滨海新区新河船厂城市设计	2014—2015年度秋季学期研究生建筑设计studio	卓信成、汪明全、蔡泽宇	硕士14级	朱文一、刘健
2	哈尔滨工业大学建筑学院	西班牙BAUM建筑设计事务所	泡泡城市	2016年哈尔滨工业大学联合设计课程教学	白思瑶、吴桐、庄昌明、Kelly Charles	本科四年级	连菲、唐家骏、Javier Caro Dominguez
3	山东建筑大学建筑城规学院	澳大利亚昆士兰理工大学	之间——济南商埠区中山公园东片综合体设计	山东建筑大学—澳大利亚昆士兰理工大学联合设计课程教学	李佩如、周继发、王天元	本科四年级	周忠凯、赵斌、Yvonne Wang、Paul Sanders
4	大连理工大学建筑与艺术学院	德国达姆施塔特大学建筑系	21世纪的图书馆——合一台中创意设计中心	建筑设计（Ⅵ）	赵玥	本科四年级	罗曜辉、范悦、于辉
5	西安建筑科技大学建筑学院	美国佛罗里达大学建筑学院	中渭桥遗址博物馆设计	2015中美联合设计课程教学	王嘉琪、范岩、O. Faiyiga、A. Urbistondo	研究生一年级	苏静、常海青、同庆楠、吴涵儒、鲁旭、Albertus Wang、Martin Gold
6	广州大学建筑与城市规划学院	香港中文大学建筑学院、英国格拉斯哥艺术学院麦金托什建筑学院	抑郁症社区康复中心	城市更新workshop——社区康复中心	陈才杰、姚志键	研究生一年级	邓毅、Ferretto Peter、蔡凌、李桔、Harry Philips
7	山东建筑大学建筑城规学院	新西兰UNITEC理工学院	泉景—泉水博物馆设计	建筑设计4	王天元	本科3年级	房文博、孔亚峰、Tony Van Raat
8	清华大学建筑学院	哥伦比亚安第斯大学、法国拉维莱特大学、意大利威尼斯建筑大学等	双面的波哥大	AIAC六校国际学生建筑设计工作坊	熊哲昆	硕士14级	程晓青、邹欢
9	中央美术学院建筑与艺术学院	瑞士卢塞恩应用科学与艺术大学建筑系	灯塔	临时展览木塔设计	李师崟、王丰	本科四年级	Dieter Geissbühler、Yves Dusseiller、Tina Unruh、Hansjürg Buchmeier、Uwe Teutsch

续表

序号	院校及院系全名	对外交流对象（国家、地区、学校及院系）	设计题目	课程名称	学生姓名	第一完成学生所在年级	指导教师姓名
10	北京建筑大学建筑学院	美国泽碧克公司	目的地的兴盛——云南抚仙湖规划建筑设计	北京建筑大学—美国泽碧克公司联合设计课程教学	翟玉琨	大学五年级	俞天琦、马衷、郭枫、田欣、Ilhan Zeybekoglu
11	西安建筑科技大学建筑学院	意大利米兰理工大学建筑学院	回溯唐的未来——唐轴线小雁塔地段城市设计	2015西安建筑科技大学—意大利米兰理工大学国际联合工作营课程教学	侯帅、Giulia Mazzuchelli、Costanza Mondani、Miriam Pozolli、赵晋、赵彬彬	研究生一年级	李昊、常海青、鲁鹏、李焜、laura Anna Pezzetti、Carlo Palazzolov
12	山东建筑大学建筑城规学院	新西兰UNITEC理工学院	织补·起航——移民社区微型功能集合体	建筑设计4	贾鹏	本科三年级	金文妍、张雅丽、Tony Burge
13	湖南大学建筑学院	捷克共和国捷克技术大学建筑学院	长沙天伦造纸厂更新改造	建筑设计A4	何磊	本科三年级	罗荩、陈翚、李旭
14	东南大学建筑学院	奥地利维也纳工业大学	南京大行宫碑亭巷旧居住区更新改造设计	建筑设计	宋文颖、白宇泓、黄迪奇、Victoria Einrach	硕士一年级	韩晓峰、葛明
15	天津大学建筑学院	意大利罗马大学建筑系	古罗马西南角旧屠宰场—陶片山—城墙区域城市设计及旧建筑加改造设计	Wandering through the History	于安然、班兴华、唐奇靓、李桃	本科四年级	卞洪滨、赵熠冬、张昕楠
16	东南大学建筑学院	加拿大不列颠哥伦比亚大学	山—海—人的相遇——温哥华华诗省沿海高速公路木休闲营地设计	建筑设计暑期实践	宗袁月、Ramona Montecillo、Robert Maggay	本科四年级	韩晓峰、屠苏南
17	华侨大学建筑学院	中国文化大学建筑及都市设计系	台北青年旅馆设计	台北华光片区青年旅馆类题设计	黄诗丹	2012级	胡璟、连旭、吴少峰
18	湖南大学建筑学院	捷克共和国捷克技术大学建筑学院	悬浮森林——滨江新城厂房改扩建之图文媒体中心	建筑设计A4	廖若微	本科三年级	严湘琦、张蔚
19	湖南大学建筑学院	斯洛文尼亚卢布尔雅那大学建筑学院	光炫之城——长沙历史街区中的社区图书馆	建筑设计A3	徐嘉韵	本科三年级	蒋甦琦、邓广

三等奖

序号	院校及院系全名	对外交流对象（国家、地区、学校及院系）	设计题目	课程名称	学生姓名	第一完成学生所在年级	指导教师姓名
1	重庆大学建筑城规学院建筑系	日本早稻田大学、中国香港大学	山地田园综合体——松阳塘后村村落更新设计	回归田园松阳：2016国际五校研究生联合设计	雷力、陈卓、万晓晓	2014级建筑学硕士生	邓蜀阳、阎波
2	哈尔滨工业大学建筑学院	英国谢菲尔德大学	复合进化论——自混沌至有机的艺术社区的自发成长	2015年哈尔滨工业大学联合设计课程教学	施雨晴、郑运潮、郭文嘉、武雪凤	12级	孟琪、梁静、Nadia Bertolino
3	同济大学建筑与城市规划学院建筑系	世界高层建筑与都市人居学会（CTBUH）、KPF建筑设计事务所	我的微纽约——位于市中央车站的垂直市场大楼	同济大学—CTBUH—KPF联合设计课程	程思、李祎喆、张谱、赵音甸	研究生一年级	谢振宇、王桢栋、谭峥
4	西北工业大学力学与土木建筑学院建筑系	中国台湾淡江大学建筑系	城嘉年华	都市市集及办公大楼设计	方帅	本科三年级	赖怡成、杨卫丽
5	大连理工大学建筑与艺术学院	中国台湾成功大学建筑系	都市驿站——垂直建筑系馆设计	建筑系馆设计	温良涵	本科四年级	颜茂仓、于辉、王时原

续表

序号	院校及院系全名	对外交流对象（国家、地区、学校及院系）	设计题目	课程名称	学生姓名	第一完成学生所在年级	指导教师姓名
6	同济大学建筑与城市规划学院建筑系	世界高层建筑与都市人居学会（CTBUH）、KPF建筑设计事务所	垂直价值激发器	同济大学—CTBUH—KPF联合设计课程	郑搴、牟娜莎、承晓宇、邬梦昊	研究生一年级	谢振宇、王桢栋、谭峥
7	重庆大学建筑城规学院建筑系	日本早稻田大学、香港大学	"织补"——基于历史文脉重塑的松阳老街更新设计	回归田园松阳：2015国际五校研究生联合设计	李忠明、马汀、王嘉睿	2014级建筑学硕士生	邓蜀阳、阎波
8	大连理工大学建筑与艺术学院	意大利米兰理工大学	旧城改造	architectural design studio 1	李东祖	本科四年级	Domenico、李冰、范悦
9	西安建筑科技大学建筑学院	意大利米兰理工大学	序列·传承——唐轴线小雁塔地段城市设计	2015西安建筑科技大学—意大利米兰理工大学国际联合工作营课程教学	卢肇松、王龙飞、李露昕、Claudia Grossi、Giulia Guida、Fillipo De Rosa	研究生一年级	李昊、常海青、鲁旭、李焜、laura Anna Pezzetti、Carlo Palazzolov
10	西安建筑科技大学建筑学院	美国佛罗里达大学建筑学院	一方天地——西安渭桥遗址区博物馆设计	2015中美联合设计课程教学	高元丰、刘佳、罗靖、Mitch clarke、Jesse Jones、Jessica Philips	研究生一年级	常海青、苏静、同庆楠、吴涵楠、鲁旭、Albertus Wang、Martin Gold
11	武汉大学城市设计学院建筑系	日本女子大学建筑系、淡江大学建筑系、逢甲大学建筑系	解析—过渡性住宅	台北京都合宅工作营	覃冢、米仓春菜、大荣桃世、阙噂桓、杨芳兰、江能煜、郭鏵湘	2012级	张睿、胡晓青、黎启国
12	南京工业大学建筑学院	美国堪萨斯大学建筑学院	年龄混合型的老年居住、健康医疗、康复社区规划与建筑设计	联合毕业设计	陈笑寒	本科五年级	蔡志昶、方遥、Hui Cai、Kent Spreckelmeyer
13	华侨大学建筑学院	澳门土地工务运输局、澳门文化局	城脉	澳门十月初五街改造与更新设计	隋路	2010级	成丽
14	哈尔滨工业大学建筑学院	西班牙BAUM建筑设计事务所	城市建筑工地预制工人住宿	2015年哈尔滨工业大学联合设计课程教学	李志斌、葛斐然、潘思傲	本科四年级	连菲、唐家骏、Javier Caro Dominguez
15	东南大学建筑学院建筑系	日本东京工业大学建筑学院	"共生"——恰园历史街区更新城市及建筑设计	中日五校联合工作坊	刘怡宁、许健、吴俊熙、Joey LIPPE、SUZUKI Aiko	研究生二年级	唐芃、葛明、奥山信一、王方戟、孙一民、胡磊
16	华南理工大学建筑学院	日本东京工业大学建筑学院	扩散	五校联合工作坊	陈碧琳	本科四年级	孙一民、李敏稚
17	哈尔滨工业大学建筑学院	英国巴斯大学	应变，随时而变	开放式研究型建筑设计	王疆、李栋梁、钟建博、赵东吉	研究生一年级	徐洪澎、吴健梅、张纹韶
18	东南大学建筑学院建筑系	日本东京工业大学日本三菱地所	南京地铁马群站周边城市设计	建筑设计IV	张宏宇、罗西	本科四年级	唐芃、沈旸、宗本顺三、惠良隆二
19	西安建筑科技大学建筑学院	意大利米兰理工大学	传承——规矩——唐轴线小雁塔地段城市设计	2015西安建筑科技大学—意大利米兰理工大学国际联合工作营课程教学	Michele Marini、Claudio livetti、Daniale Delgrosso、廖枢丹、孙雅雯、邢泽坤	研究生一年级	李昊、常海青、鲁旭、李焜、laura Anna Pezzetti、Carlo Palazzolov
20	西安建筑科技大学建筑学院	意大利米兰理工大学	行走丝绸路之上	2015西安建筑科技大学—意大利米兰理工大学国际联合工作营课程教学	尹锐莹、Alberto Malabarba、Francesco Busnelli、路冠丞、张雅楠、张婧琪	研究生一年级	李昊、常海青、鲁旭、李焜、laura Anna Pezzetti、Carlo Palazzolov
21	山东建筑大学建筑城规学院	澳大利亚昆士兰理工大学	步行城市——布里斯班韦斯滕德城市设计	山东建筑大学—澳大利亚昆士兰理工大学联合设计课程教学	朱轩毅、贾慧、殷子君、党常顾	本科四年级	张克强、任震、Yvonne Wang、Tony Van Raat
22	山东建筑大学建筑城规学院	新西兰UNITEC理工学院	隙光——意大利普拉托考古遗址博物馆设计	建筑设计3	仲文	本科三年级	慕启鹏、孔亚暐、Tony Burge

续表

序号	院校及院系全名	对外交流对象（国家、地区、学校及院系）	设计题目	课程名称	学生姓名	第一完成学生所在年级	指导教师姓名
23	太原理工大学建筑与土木工程学院建筑系	荷兰Karel事务所	越陌度阡	历史街区更新工作坊	郝志伟、罗艾婧	本科四年级	徐强、高静、董艳平、Karel Nieuwland
24	南京大学建筑与城市规划学院	英国剑桥大学建筑系	互动——剑桥大学学制下公共空间的分析与设计	南京大学—剑桥大学联合毕业设计	黎乐源	本科四年级	窦平平、Ingrid Schröder
25	浙江大学建筑工程学院建筑系	西澳大利亚大学、澳大利亚城市设计研究中心	万花城——人与生态的城市设计	专题化设计	严子君	本科四年级	Joerg Baumeister、Daniela A. Ottmann、吴越、王卡

入围奖

序号	院校及院系全名	对外交流对象（国家、地区、学校及院系）	设计题目	课程名称	学生姓名	第一完成学生所在年级	指导教师姓名
1	北方工业大学建筑与艺术学院	美国加利福尼亚州立理工大学伯莫纳分校	种植·培养·建造	北方工业大学—加利福尼亚州立理工大学伯莫纳分校联合设计	张雅琪、林悦、王悠然、Mutsawashe Chipfumbu、Cyndi Feris、Umurinzi Serge、Niyoyita Joseline、Zenas Guo、Alan Hu、Lorena Jauregui、Arturo Ortuno、Julianne Pineda、Eddy Solis、Joseph Jamoralin、Hakizimana Dusenga Yvette	本科四年级	秦柯、Irma Ramirez、Courtney Knapp、张勃、安平
2	北方工业大学建筑与艺术学院	美国加利福尼亚州立理工大学伯莫纳分校	地球村	北方工业大学—加利福尼亚州立理工大学伯莫纳分校联合设计	薛皓硕、陈婉钰、苏伊莎、艾克拜尔、Uwimana Lydia、Nurlan Babayev、Tinashe Justin Machamire、Emily Williams、Houra Khani、Sergio Gutierre、Juan Galvan、Renzo Pali、Pan Chunguang、Ihumere Irma、Mukazera Marie Christelle	本科四年级	秦柯、Irma Ramirez、Courtney Knapp、张勃、安平
3	北京建筑大学建筑学院	德国柏林工业大学	保福寺地区城市更新项目	柏林工业大学—中国建筑设计研究院北京建筑大学课程教学	王天娇、徐丹、王任书、耿云楠	14级研究生	马英、欧阳文、景泉、李静威
4	北京建筑大学建筑学院	德国柏林工业大学	保福寺地区城市更新项目	柏林工业大学—中国建筑设计研究院北京建筑大学课程教学	杜京伦、魏立志、张明	14级研究生	王佐、马英、欧阳文、景泉、李静威
5	北京建筑大学建筑学院	德国柏林工业大学	保福寺地区城市更新项目	柏林工业大学—中国建筑设计研究院北京建筑大学课程教学	李取奇、王晓健、牛亚庆	14级研究生	晁军、王佐、马英、景泉、李静威
6	大连理工大学建筑与艺术学院	德国达姆施塔特大学建筑系	博登湖畔市政厅设计	建筑设计	尚书	本科五年级	Meinrad Morger、王时原、吴亮
7	大连理工大学建筑与艺术学院	意大利米兰理工大学建筑系	青年旅馆设计	建筑设计（七）	崔培睿	本科五年级	杜方中、吴亮、李冰
8	大连理工大学建筑与艺术学院	韩国高丽大学工程学院建筑系	和平市场改造计划	建筑设计Ⅵ	徐佳臻	本科五年级	Fabio Dacarro、李冰、吴亮

续表

序号	院校及院系全名	对外交流对象（国家、地区、学校及院系）	设计题目	课程名称	学生姓名	第一完成学生所在年级	指导教师姓名
9	东南大学建筑学院建筑系	日本东京工业大学	社区建筑——东南大学书院设计	建筑设计IV	沈忱、隋明明	本科四年级	葛明、奥山信一
10	东南大学建筑学院建筑系	中国香港大学建筑学院	住区设计	建筑设计IV	李鸿渐	本科四年级	贾倍思
11	东南大学建筑学院建筑系	美国艾奥瓦州立大学	波士顿新区音乐广场设计	毕业设计	马斯文	本科五年级	鲍莉
12	南京工业大学建筑学院	美国堪萨斯大学建筑学院	年龄混合型的老年居住、健康、医疗、康复社区规划与建筑设计	联合毕业设计	梁末、季鑫	本科五年级	蔡志昶、方遥、Hui Cai、Kent Spreckelmeyer
13	西安建筑科技大学建筑学院	意大利米兰理工大学建筑学院	转变·演绎——唐轴线朱雀门顺城巷地段规划与城市设计	2015西安建筑科技大学—意大利米兰理工大学国际联合工作营课题教学	陈哲怡、崔焌曼、方坚、徐娉、Alessandra Guizzi, Giulia Dagheti、Greta Bosio、武琼	研究生一年级	李昊、常海青、鲁旭、Laura Anna Pezzetti, Carlo Palazzolov
14	西安建筑科技大学建筑学院	美国佛罗里达大学都市设计学院	行·走·延续——西安渭桥遗址区博物馆设计	2015中美联合设计课程教学	陈哲怡、崔焌曼、Wendy Stradley、Kinga Pabjan	研究生一年级	常海青、苏静、吴涵儒、鲁旭、Albertus Wang、Martin Gold
15	华南理工大学建筑学院	美国加利福尼亚大学伯克利分校建筑学院	啤酒厂改造	SCUT—UCB联合工作坊琶洲城市设计	梁雅晴、李越宜、司竞、吴巨荣	本科四年级	孙一民、Peter Bosslmann、苏平、王璐、周毅刚、李敏稚
16	华南理工大学建筑学院	意大利都灵理工大学建筑学院	T.I.T工业创意园城市设计策略探究	五年级导师制	郭院、洪梦扬、李泳妍、倪安琪	本科五年级	孙一民、Francesca Frassoldati、李敏稚
17	华侨大学建筑学院	中国文化大学建筑及都市设计学院	活力台北·青年文创SOHO集合住宅设计		王欣远	2012级	胡璟、连旭、吴少峰
18	华侨大学建筑学院	澳门土地工务运输局、澳门文化局	点·聚生活	澳门十月初五街改造与更新设计	叶子颖	2010级	费迎庆
19	广州大学建筑与城市规划学院	香港中文大学建筑学院、英国格拉斯哥艺术学院麦金托什建筑学院	社区活动中心	城市更新workshop——社区康复中心	李浩、陈彬新	研究生一年级	蔡凌、Ferretto Peter、邓毅、李桔、Harry Philips
20	广州大学建筑与城市规划学院	香港中文大学建筑学院、英国格拉斯哥艺术学院麦金托什建筑学院	自闭症社区康复中心	城市更新workshop——社区康复中心	胡彬、黄智尚	研究生二年级	蔡凌、Ferretto Peter、邓毅、李桔、Harry Philips
21	广州大学建筑与城市规划学院	英国格拉斯哥哥艺术学院麦金托什建筑学院	城中村中的新市场	洗墩村更新城市设计workshop	刘健、胡彬、覃可妍、劳佩珊	本科四年级	邓毅、Alan Hooper、李桔、路尔提
22	山东建筑大学建筑城规学院	新西兰UNITEC理工学院	伊特鲁里亚的复兴——普拉托Gonfient地区遗址公园设计	山东建筑大学—新西兰奥克兰理工学院联合设计课程教学	于陈晨、张渠、张明辉	本科四年级	常玮、刘建军、Su Bin、Luca Piantini
23	合肥工业大学建筑与艺术学院	逢甲大学建筑系	缓缓归——都市集合住宅设计	建筑设计	张雅倩	本科三年级	许学礼
24	天津大学建筑学院	意大利罗马大学建筑学院	唤醒历史——古罗马旧屠宰场—陶片山—城墙区域城市设计及旧建筑加建设计	古罗马西南角旧屠宰场—陶片山—城墙区域城市设计及旧建筑加建设计	李文爽、崔家瑞、李宗泽	本科四年级	张昕楠、卜洪滨、赵娜冬
25	同济大学建筑与城市规划学院建筑系	世界高层建筑与都市人居学会(CTBUH)、KPF建筑设计事务所	垂直终点站	同济大学—CTBUH—KPF联合设计课程	李鹭、刘晓宇、陆伊昀、杨之媛	研究生一年级	谢振宇、王桢栋、谭峥
26	福建工程学院建筑与城乡规划学院	台湾科技大学	明眸心塾——基于新理念下的视障教育学校设计	毕业设计	陈星	本科五年级	余志红、高小情